Modern EMC Analysis Techniques

Volume II: Models and Applications

Modern EMC Analysis Techniques. Volume II: Models and Applications
Nikolaos V. Kantartzis and Theodoros D. Tsiboukis

ISBN: 978-3-031-00578-7   paperback
ISBN: 978-3-031-01706-3   ebook

DOI: 10.1007/978-3-031-01706-3

A Publication in the Springer series

SYNTHESIS LECTURES ON COMPUTATIONAL ELECTROMAGNETICS # 22

Lecture #22

Series Editor: Constantine A. Balanis, Arizona State University

Series ISSN
ISSN 1932-1252   print
ISSN 1932-1716   electronic

Modern EMC Analysis Techniques
Volume II: Models and Applications

Nikolaos V. Kantartzis and Theodoros D. Tsiboukis
Aristotle University of Thessaloniki

SYNTHESIS LECTURES ON COMPUTATIONAL ELECTROMAGNETICS # 22

ABSTRACT

The objective of this two-volume book is the systematic and comprehensive description of the most competitive time-domain computational methods for the efficient modeling and accurate solution of modern real-world EMC problems. Intended to be self-contained, it performs a detailed presentation of all well-known algorithms, elucidating on their merits or weaknesses, and accompanies the theoretical content with a variety of applications. Outlining the present volume, numerical investigations delve into printed circuit boards, monolithic microwave integrated circuits, radio frequency microelectromechanical systems as well as to the critical issues of electromagnetic interference, immunity, shielding, and signal integrity. Biomedical problems and EMC test facility characterizations are also thoroughly covered by means of diverse time-domain models and accurate implementations. Furthermore, the analysis covers the case of large-scale applications and electrostatic discharge problems, while special attention is drawn to the impact of contemporary materials in the EMC world, such as double negative metamaterials, bi-isotropic media, and several others.

KEYWORDS

electromagnetic compatibility (EMC), time-domain methods, computational electromagnetics

Preface

In writing a book on contemporary electromagnetic compatibility (EMC) analysis techniques, the authors are well aware that they are delving into a field of challenging innovations pertinent to a large readership ranging from students and academics to engineers and seasoned professionals. Nowadays, EMC technology is more pervasive than ever in many educational, social, industrial, and commercial sectors. Essentially, the widely recognized significance of EMC problems and applications has turned the interest of the scientific community toward their in-depth investigation, with an emphasis on the simulation of the relevant electromagnetic phenomena via highly advanced time-domain methodologies. This is exactly the aim of the two-volume book, which basically reflects the outgrowth of a systematic research performed by the authors for almost a 10-year period in the area of EMC/electromagnetic interference (EMI) measurement and modeling. Intended to be self-contained from the computational perspective, the book performs a detailed presentation of most time-domain algorithms, elucidating on their merits or weaknesses, and accompanies the theoretical content with a variety of real-world applications. Thus, having acquired the necessary evidence for every numerical approach, the reader is then free to decide on the best possible scheme for his/her requirements.

Outlining the second volume, dealing with the numerical solution of modern EMC problems and comparisons with measurement data, Chapter 1 gives a short introduction on the different types of applications and discusses which time-domain techniques can be utilized for their treatment. Hence, proceeding to the central task of this book, Chapter 2 investigates the topic of printed circuit boards and after some short theoretical analysis, addresses diverse power bus configurations, decoupling devices, and switching interconnects. Furthermore, special sections are devoted to the examination of monolithic microwave integrated circuits and radio frequency microelectromechanical systems. Next, Chapter 2 handles the critical issues of EMI, immunity, shielding, and signal integrity that are proven very instructive during the design of an EMC device. A variety of structures is pursued, such as antennas, wireless local area networks, slot cavities, and multiport waveguides. Biomedical problems and human exposure to electromagnetic fields, in particular, are the subject of Chapter 4. Herein, the concept of dispersive time-domain algorithms is comprehensively clarified and several ways to compute the specific absorption rate and thermal gauges are discussed. The chapter contains an elaborate set of results from exposure to realistic cellular phones, antennas, and wireless radiators. Moreover, the characterization of EMC test facilities is performed in Chapter 5. To this aim, a large list of already constructed anechoic/semianechoic, reverberation chambers and TEM cells is explored, whereas new designs are thoroughly verified according to international standards. Chapter 6, on the other hand, suggests alternative means for the affordable manipulation of large-scale problems and electrostatic discharge cases, whereas Chapter 7 closes the book with an extensive reference to contemporary materials, such as the double-negative metamaterials, bi-isotropic media, nanotechnology layouts, and photonic crystal arrangements.

Finally, the authors would like to thank Dr. T. T. Zygiridis for his thorough proofreading and valuable suggestions during the preparation of the manuscripts. Above all, they do anticipate that the theoretical formulations and numerical results provided in both volumes will inspire the reader to expand its material beyond the prescribed limits and efficiently reimburse the lack of a microwave laboratory, which is rather expensive to construct, but so indispensable in teaching.

Thessaloniki   **N. V. Kantartzis**
December 2007   **T. D. Tsiboukis**

Contents

CHAPTER 1

Introduction

1.1 THE PERCEPTION OF MODELING IN THE AREA OF ELECTROMAGNETIC COMPATIBILITY

The process of constructing the appropriate model of an electrical system or a physical phenomenon is a particularly instructive means for an engineer, because it offers practical simulations for a large assortment of initial conditions, excitations, or layouts. In fact, the notions of modeling have always attracted the interest of researchers, who have long attempted to build concrete equivalents of—sometimes abstract—mechanisms that could not be easily tested, measured, or even felt by the senses. Therefore, the first models appeared when the association between mathematics and experimental observation of natural procedures became feasible. In the broad area of electromagnetics, one of the earliest models has been the representation of the magnetic effect created by an electric current using the force lines of the magnetic field. Basically, the majority of such outgrowths are founded on Maxwell's equations, which constitute the most general tool for the description of physical interactions. Combining the accumulated knowledge together with a sound understanding of nature, these laws have given us the possibility of delving into more complicated systems and determine the necessary parameters for their analysis in a reasonable time and accuracy level. However, *no* model is deemed consistent if it cannot be thoroughly validated or equivalently certified by the pertinent reference results or measurement data and hence supply global interpretations.

Bearing in mind the above, the modeling procedure in the field of *electromagnetic compatibility (EMC)* signifies the establishment of a relationship between a source or interference and its effect, such as the response of a system [1, 2]. Apart from the conventional methods, based on circuit theory, to achieve this objective the most contemporary one is to numerically represent the problem (either in frequency or time domain) by means of formal solutions to Maxwell's laws and the necessary boundary conditions. Actually, the term EMC indicates the competence of an electrical apparatus to operate in its specified electromagnetic environment within balanced thresholds of safety and efficiency, without suffering from or triggering intolerable degradation as an upshot of unwanted disruptions. Because EMC is strongly connected to electrical noise suppression, its initial advent may be associated to the use of very low currents in practical applications. Since then, the fabrication of electronic components, increasingly sensitive to lower energies, has aroused the need to render these devices durable in disturbed environments and high percentages of *electromagnetic interference (EMI)*. To this aim, all potential models, and especially the numerical ones, should carefully consider the conservation of energy, the causality of the product's response, the utilized excitation waveform as well as the low- or high-frequency behavior of spectral modes generated by unwanted oscillations at different media interfaces. Nevertheless, before resolving the suitable methodology for the attainment of the prior issues, it is mandatory to categorize the main types of EMC problems in order to isolate their chief traits and peculiarities.

1.2   GENERAL CLASSES OF MODERN EMC PROBLEMS

The large variety of EMC applications originates from the corresponding number of different EMI sources [1]. Typically, in developing a numerical model, the first stage is to classify the propagation path between the EMI source and the object. A generic, yet fairly descriptive, categorization leads to two distinct types: the natural and the man-made noise, although there are some alternative criteria that divide the sources into continuous wave and transient ones. In this manner, the researcher is able to recognize the main characteristics of its problem and then proceed to its discretization. Nevertheless, not all physical phenomena can be adequately modeled, thus raising questions of accuracy and efficiency. At the moment, the most significant limitation in EMC modeling is the complexity of the real object or arrangement. For instance, even though it is feasible to precisely simulate the coupling among an electromagnetic field and several parallel transmission lines, it is not possible to produce acceptable results in the case of cables comprising nonshielded conductors and coaxial lines toward a complicated routing inside a vehicle or a building. Because the stipulations for enhanced quality and high-end competences are still under rapid evolution, it becomes obvious that computational models must be able to predict all system-level responses to external and internal EMI, evaluate the behavior of EMC protection measures, and suggest supplementary design directions.

The fact that EMC/EMI interactions are recognized as a very important part of the manufacturing industry is justified by the serious theoretical and experimental research conducted nowadays. Indeed, the constant progress of electrical and electronics engineering has led to impressive advances in many scientific areas [2]. Consequently, the main categories of EMC applications include printed circuit boards, modern radiators, radio frequency microwave devices or waveguiding setups, monolithic circuitry and packaging interconnects, communication systems, shielding sheets, immunity testers, bioelectromagnetic systems, anechoic or reverberating chambers, large-scale structures, electrostatic discharge configurations, and the latest double-negative metamaterial, chiral, nanotechnology, and photonic crystal implementations. Taking into account these classes of problems, it is indisputable that their numerical investigation is not only a matter of programming or computer power. On the contrary, certain formulations with a robust mathematical profile are required. This is exactly where time-domain *compu-*

tational electromagnetics (CEM) can give a promising assistance. After a period of continuous growth, CEM has a number of advanced methodologies that are capable of dealing with real-world problems. The majority of these algorithms keep on improving via some truly intuitive ideas, through which they try to subdue rudimentary weaknesses or alleviate stability defects.

1.3 MOTIVES FOR SELECTING A TIME-DOMAIN TECHNIQUE

Time-domain methods use the *direct* solution of Maxwell's curl equations by sampling electric and magnetic field vectors in a temporal window via the pertinent stability conditions [3–5]. Their update concept stems from an explicit integration scheme for the computation of every unknown component without the necessity of solving any system of equations, unlike their frequency-domain counterparts. Furthermore, spatial discretization assigns quantities on structured uncollocated grids of subwavelength resolutions designed to effectively handle the frequency content of the excitation according to all consistency rules. To the aforesaid features, one should count the relatively easy and concurrently intelligent manipulation of material interfaces, outer boundaries, and periodic structures. It is also worth noting the broadband operation as well as the incorporation of diverse modular schemes as amendments to specific modeling difficulties. Some nontrivial shortcomings, such as the artificial lattice reflection errors or the "staircase" spatial approximation, may be circumvented via combined algorithms and hybrid formulations. This is particularly useful for objects of large electrical size, strenuous shape, or cross section, and multifrequency cases. Principally, provided the interleaved nature and the rigorous stencil-control logic of time-domain methods, Maxwell's laws can be safely transformed to algebraic links between field vectors and therefore yield some really challenging EMC models.

1.4 TIME-DOMAIN METHODS AMENABLE TO EMC ANALYSIS[1]

Computational modeling of electromagnetic problems in the time domain has matured a lot during the past two decades. Therefore, apart from the already popular techniques for EMC applications, a number of efficient algorithms have been introduced and gained a remarkable appreciation. Our brief reference to the formulations that will be used throughout this book, starts, of course, from the finite-difference time-domain (FDTD) algorithm [6–8], the transmission line matrix/modeling method (TLM) [9–11], and the finite integration technique (FIT) [12]. The first one constructs a

---

[1]For a detailed theoretical description of all time-domain numerical methods, cited in this book, through various mathematical formulations and helpful examples, the reader is referred to its first volume: *"Modern EMC Analysis Techniques: Time-Domain Computational Schemes"* of the same authors and publisher.

staggered lattice in space–time and places the unknown field quantities according to a certain nodal pattern. Thus, for spatial sampling, electric components, which are located at cell edges, are circulated by four magnetic components, placed at cell faces and vice versa. Similar to its implementation is the temporal integration, which complies with the rules of the robust leapfrog scheme. On the other hand, the second method computes electromagnetic waves through scattered and propagating pulses, positioned at predetermined points of the grid. This concept represents field components via a set of planes with the wave amplitudes perpendicular to them in order to exploit the advantages of a network. Finally, the third approach exploits the integral form of Maxwell's equations and uses six components of electric and magnetic voltages/fluxes on a dual energy-conserving mesh. Analytically, voltages are linked to contours, whereas fluxes are linked to surfaces, so enabling the construction of a matrix system, pertinent for curvatures and fine details.

Although the preceding formulations aggregate a lot of advantages, they have weaknesses or remain vulnerable to the traditional inherent discretization threats. For these deficiencies to be mitigated, various explicit algorithms and hybridizations have been proposed, such as the finite-element time-domain (FETD) [13, 14] method and the finite-volume time-domain (FVTD) [15] technique for the treatment of arbitrarily curved surfaces and general coordinate systems. Moreover, one should mention the multiresolution time-domain algorithm (MRTD) [16] and the pseudospectral time-domain (PSTD) [17] method for the manipulation of EMC structures with high periodicity, whereas the alternating-direction implicit (ADI) FDTD approach [18, 19] allows the choice of temporal increments beyond the Courant stability limit, without generating any oscillatory or exponentially growing wave modes. In addition, the suppression of the undesired and often detrimental dispersion, anisotropy, and dissipation errors may be accomplished through the nonstandard FDTD method or other higher-order schemes [20]. Although not exhaustive, the above list contains the most important and robust methodologies for the computational solution of realistic EMC problems. Being familiar to their key merits, the reader is prompted to view numerical simulations as a test bench for their accuracy, convergence as well as consistency and the most critical: for the correction of their potential shortcomings.

REFERENCES

1. C. R. Paul, *Introduction to Electromagnetic Compatibility*, 2nd ed. New York: Wiley-Interscience, 2006.
2. M. I. Montrose and E. D. Nakauchi, *Testing for EMC Compliance – Approaches and Techniques*. New York: IEEE Press and Wiley-Interscience, 2004.
3. A. F. Peterson, S. L. Ray, and R. Mittra, *Computational Methods for Electromagnetics*. Piscataway, NJ: IEEE Press and Oxford University Press, 1998.
4. D. B. Davidson, *Computational Electromagnetics for RF and Microwave Engineering*. Cambridge, UK: Cambridge University Press, 2005.

5. H.-D. Brüns, C. Schuster, and H. Singer, "Numerical electromagnetic field analysis for EMC problems," *IEEE Trans. Electromagn. Compat.*, vol. 49, no. 2, pp. 253–262, May 2007.

6. K. S. Yee, "Numerical solution of initial boundary value problems involving Maxwell's equations in isotropic media," *IEEE Trans. Antennas Propagat.*, vol. AP-14, no. 3, pp. 302–307, May 1966.

7. K. S. Kunz and R. J. Luebbers, *The Finite Difference Time Domain Method for Electromagnetics*. Boca Raton, FL: CRC Press, 1993.

8. A. Taflove and S. C. Hagness, *Computational Electrodynamics: The Finite-Difference Time-Domain Method*, 3rd ed. Norwood, MA: Artech House, 2005.

9. W. J. R. Hoefer, "The transmission line matrix (TLM) method," in *Numerical Techniques for Microwave and Millimeter Wave Passive Structures*, T. Itoh, Ed. New York: John Wiley and Sons, 1989, pp. 496–451.

10. M. Krumpholz and P. Russer, "A field theoretical derivation of TLM," *IEEE Trans. Microwave Theory Tech.*, vol. 42, no. 9, pp. 1660–1668, Sept. 1994. doi:10.1109/22.310559

11. C. Christopoulos, *The Transmission-Line Modeling Method: TLM*. New York: IEEE Press, 1995.

12. T. Weiland, "A discretization method for the solution of Maxwell's equations for six-component fields," *Electron. Commun. (AEÜ)*, vol. 31, no. 3, pp. 116–120, 1977.

13. J.-F. Lee, R. Lee, A. C. Cangellaris, "Time-domain finite-element methods," *IEEE Trans. Antennas Propagat.*, vol. 45, no. 3, pp. 430–442, Mar. 1997.

14. J.-M. Jin, *The Finite Element Method in Electromagnetics*, 2nd ed. New York: John Wiley and Sons, 2002.

15. R. Holland, V. P. Cable, and L. C. Wilson, "Finite-volume time-domain (FVTD) techniques for EM scattering," *IEEE Trans. Electromagn. Compat.*, vol. 33, no. 4, pp. 281–294, Nov. 1991. doi:10.1109/15.99109

16. M. Krumpholz and L. P. B. Katehi, "MRTD: New time-domain schemes based on multiresolution analysis," *IEEE Trans. Microwave. Theory Tech.*, vol. 44, no. 4, pp. 555–571, Apr. 1999.

17. Q. H. Liu, "The PSTD algorithm: A time-domain method requiring only two cells per wavelength," *Microwave. Opt. Technol. Lett.*, vol. 15, no. 3, pp. 158–165, 1997. doi:10.1002/(SICI)1098-2760(19970620)15:3<158::AID-MOP11>3.0.CO;2-3

18. T. Namiki, "A new FDTD algorithm based on ADI method," *IEEE Trans. Microwave. Theory Tech.*, vol. 47, no. 10, pp. 2003–2007, Oct. 1999.

19. F. Zheng, Z. Chen, and J. Zhang, "A finite-difference time-domain method without the Courant stability conditions," *IEEE Microwave. Guided Wave Lett.*, vol. 9, no. 11, pp. 441–443, Nov. 1999. doi:10.1109/75.808026

20. N. V. Kantartzis and T. D. Tsiboukis, *Higher-Order FDTD Schemes for Waveguide and Antenna Structures*, San Rafael, CA: Morgan & Claypool Publishers, 2006.

CHAPTER 2

Printed Circuit Boards in EMC Structures

2.1   INTRODUCTION

Printed circuit boards (PCBs) constitute a rapidly evolving field of modern microwave engineering with a variety of promising research directions. Combining their high design flexibility with the properties of innovative geometric conceptions, they soon became the primary vehicle for the realization of many theoretical findings. However, their proper function is strongly related to the environment or the host structure they are built in, and therefore any attempt to identify possible violations of the international EMC standards before the first production, is deemed crucial. Toward this objective, the merits of time-domain computer modeling can offer viable solutions to the professional expert by revealing potential interference sources and undesired coupling mechanisms. Actually, through a well-established numerical algorithm an EMC engineer can rapidly handle an enormous amount of information and obtain the correct parameters for the final fabrication process. In this context, the present chapter sheds light to the analysis of diverse realistic PCB arrangements by means of different time-domain methodologies and hybrid schemes. Amid the problems addressed are DC-power bus devices, decoupling components, switching interconnects with multiple vias, monolithic microwave integrated circuits, and microelectromechanical systems (MEMS). From the results, often compared with available measurement or reference data, it is readily derived that time-domain modeling can serve as a reliable design instrument for complicated PCB structures in many EMC applications.

2.2   MODELING TECHNIQUES FOR PCB PROBLEMS

Before proceeding to the study of several contemporary PCB setups and their immunity profile, it is deemed instructive to provide a short, yet informative, description of the basic time-domain numerical formulations used in this chapter. These are the FDTD algorithm, the TLM method, and the PSTD technique. The theoretical presentation focuses on the most significant realization details from an EMC point of view and gives the final 3-D form of the explicit field-update equations to be implemented in the respective programming codes.

2.2.1   The Finite-Difference Time-Domain Algorithm

The FDTD method is, unquestionably, the dominant time-domain technique in the area of PCB modeling [1–3]. Combining the simplicity and mathematical robustness with the systematic extraction of meaningful wideband results, it offers the opportunity for rigorous simulations without the need of adopting nonphysical assumptions. Furthermore, its prominent versatility allows for fairly efficient hybridizations with other schemes, which, conditional on the problem under inspection, can lead to drastic improvements.

Consider an $f(x, y, z, t)$ function with continuous derivatives in a 3-D space Ω, subdivided into cells of Δx, Δy, and Δz spatial increments. All discretizations follow the notation of

$$f(x, y, z, t) = f(i\Delta x, j\Delta y, k\Delta z, n\Delta t) = f|_{i,j,k}^{n}$$

$$\left.\frac{\partial f}{\partial z}\right|_{(x, y, z, t)} = \left.\frac{\partial f}{\partial z}\right|_{(i\Delta x, j\Delta y, k\Delta z, n\Delta t)} = \left.\frac{\partial f}{\partial z}\right|_{i,j,k}^{n},$$

(2.1)

with Δt the analogous temporal increment. For a sufficiently small Δz and a certain instant $t = n\Delta t$, derivative $\partial f/\partial z$ is approximated by a central finite-difference scheme, as

$$\left.\frac{\partial f}{\partial z}\right|_{(x, y, z, t)} = \frac{f\left(x, y, z + \frac{\Delta z}{2}, t\right) - f\left(x, y, z - \frac{\Delta z}{2}, t\right)}{\Delta z} \Rightarrow$$

(2.2)

$$\left.\frac{\partial f}{\partial z}\right|_{i,j,k}^{n} = \frac{f|_{i,j,k+1/2}^{n} - f|_{i,j,k-1/2}^{n}}{\Delta z} + O\left[(\Delta z)^2\right]$$

In the above, the $\pm 1/2$ shift establishes the spatial staggering of the method, because it interleaves electric **E** and magnetic **H** field intensity vectors. The process is completed by evaluating the time partial derivatives in terms of the same central-difference expressions. Hence, for a fixed grid node, it is acquired

$$\left.\frac{\partial f}{\partial t}\right|_{(x, y, z, t)} = \frac{f\left(x, y, z, t + \frac{\Delta t}{2}\right) - f\left(x, y, z, t - \frac{\Delta t}{2}\right)}{\Delta t} \Rightarrow$$

(2.3)

$$\left.\frac{\partial f}{\partial t}\right|_{i,j,k}^{n} = \frac{f|_{i,j,k}^{n+1/2} - f|_{i,j,k}^{n-1/2}}{\Delta t} + O\left[(\Delta t)^2\right],$$

where superscripts $n \pm 1/2$ launch the well-known leapfrog integration process. In essence, the FDTD algorithm deals with the solution of initial-boundary value electromagnetic problems. In fact, electric variables, located in the middle of cell edges, are surrounded by four circulating magnetic variables. Similarly, magnetic quantities, placed at the centers of cell faces, are surrounded by four circulating electric quantities. This setup is critical, because it complies with the general rules of optimal discretization and yields a straightforward contour allocation mapping for Ampère's and Faraday's integral laws via dually interlinked current flux loops. Therefore, in the case of a domain occupied by media with a constant variation of their properties, Maxwell's curl equations conclude to the subsequent time-advancing expressions

$$
\begin{aligned}
E_x|_{i,j,k}^{n+1} = ep_{i,j,k}^{E_x} E_x|_{i,j,k}^{n} + eq_{i,j,k}^{E_x,\Delta y} \left[H_z|_{i,j+1/2,k}^{n+1/2} - H_z|_{i,j-1/2,k}^{n+1/2} \right] \\
- eq_{i,j,k}^{E_x,\Delta z} \left[H_y|_{i,j,k+1/2}^{n+1/2} - H_y|_{i,j,k-1/2}^{n+1/2} \right],
\end{aligned}
\tag{2.4}
$$

$$
\begin{aligned}
E_y|_{i,j,k}^{n+1} = ep_{i,j,k}^{E_y} E_y|_{i,j,k}^{n} + eq_{i,j,k}^{E_y,\Delta z} \left[H_x|_{i,j,k+1/2}^{n+1/2} - H_x|_{i,j,k-1/2}^{n+1/2} \right] \\
- eq_{i,j,k}^{E_y,\Delta x} \left[H_z|_{i+1/2,j,k}^{n+1/2} - H_z|_{i-1/2,j,k}^{n+1/2} \right],
\end{aligned}
\tag{2.5}
$$

$$
\begin{aligned}
E_z|_{i,j,k}^{n+1} = ep_{i,j,k}^{E_z} E_z|_{i,j,k}^{n} + eq_{i,j,k}^{E_z,\Delta x} \left[H_y|_{i+1/2,j,k}^{n+1/2} - H_y|_{i-1/2,j,k}^{n+1/2} \right] \\
- eq_{i,j,k}^{E_z,\Delta y} \left[H_x|_{i,j+1/2,k}^{n+1/2} - H_x|_{i,j-1/2,k}^{n+1/2} \right],
\end{aligned}
\tag{2.6}
$$

$$
\begin{aligned}
H_x|_{i,j,k}^{n+1/2} = hp_{i,j,k}^{H_x} H_x|_{i,j,k}^{n-1/2} + hq_{i,j,k}^{H_x,\Delta z} \left[E_y|_{i,j,k+1/2}^{n} - E_y|_{i,j,k-1/2}^{n} \right] \\
- hq_{i,j,k}^{H_x,\Delta y} \left[E_z|_{i,j+1/2,k}^{n} - E_z|_{i,j-1/2,k}^{n} \right],
\end{aligned}
\tag{2.7}
$$

$$
\begin{aligned}
H_y|_{i,j,k}^{n+1/2} = hp_{i,j,k}^{H_y} H_y|_{i,j,k}^{n-1/2} + hq_{i,j,k}^{H_y,\Delta x} \left[E_z|_{i+1/2,j,k}^{n} - E_z|_{i-1/2,j,k}^{n} \right] \\
- hq_{i,j,k}^{H_y,\Delta z} \left[E|_{i,j,k+1/2}^{n} - E_x|_{i,j,k-1/2}^{n} \right],
\end{aligned}
\tag{2.8}
$$

$$
\begin{aligned}
H_z|_{i,j,k}^{n+1/2} = hp_{i,j,k}^{H_z} H_z|_{i,j,k}^{n-1/2} + hq_{i,j,k}^{H_z,\Delta y} \left[E_x|_{i,j+1/2,k}^{n} - E_x|_{i,j-1/2,k}^{n} \right] \\
- hq_{i,j,k}^{H_z,\Delta x} \left[E_y|_{i+1/2,j,k}^{n} - E_y|_{i-1/2,j,k}^{n} \right],
\end{aligned}
\tag{2.9}
$$

with $ep_{i,j,k}^f = \dfrac{2\varepsilon_{i,j,k} - \sigma_{i,j,k}\Delta t}{2\varepsilon_{i,j,k} + \sigma_{i,j,k}\Delta t}$, $eq_{i,j,k}^{f,\Delta} = \dfrac{2\Delta t}{\Delta\left(2\varepsilon_{i,j,k} + \sigma_{i,j,k}\Delta t\right)}$,   for $\begin{cases} f = E_x, E_y, E_z \\ \Delta = \Delta x, \Delta y, \Delta z \end{cases}$,

$$\hspace{6cm}\text{(2.10)}$$

$hp_{i,j,k}^f = \dfrac{2\mu_{i,j,k} - \rho'_{i,j,k}\Delta t}{2\mu_{i,j,k} + \rho'_{i,j,k}\Delta t}$, $hq_{i,j,k}^{f,\Delta} = \dfrac{2\Delta t}{\Delta\left(2\mu_{i,j,k} + \rho'_{i,j,k}\Delta t\right)}$,   for $\begin{cases} f = H_x, H_y, H_z \\ \Delta = \Delta x, \Delta y, \Delta z \end{cases}$

coefficients referring to each electric and magnetic component at point (i, j, k). In particular, if the domain contains media with diverse constitutive properties, (2.4)–(2.9) can be launched in a more compact way. This is conducted through the suitable number (one to six) of integer arrays depending on the **E** and **H** quantities that interact with these materials. For illustration, the update of E_x and H_y becomes

$$E_x\big|_{i,j,k}^{n+1} = \text{ep}[l_e] \cdot E_x\big|_{i,j,k}^{n} + \text{eqy}[l_e] \cdot \left[H_z\big|_{i,j+1/2,k}^{n+1/2} - H_z\big|_{i,j-1/2,k}^{n+1/2} \right]$$
$$- \text{eqz}[l_e] \cdot \left[H_y\big|_{i,j,k+1/2}^{n+1/2} - H_y\big|_{i,j,k-1/2}^{n+1/2} \right], \hspace{2cm}\text{(2.11)}$$

$$H_y\big|_{i,j,k}^{n+1/2} = \text{hp}[l_h] \cdot H_y\big|_{i,j,k}^{n-1/2} + \text{hqx}[l_h] \cdot \left[E_z\big|_{i+1/2,j,k}^{n} - E_z\big|_{i-1/2,j,k}^{n} \right]$$
$$- \text{hqz}[l_h] \cdot \left[E_x\big|_{i,j,k+1/2}^{n} - E_x\big|_{i,j,k-1/2}^{n} \right], \hspace{2cm}\text{(2.12)}$$

where $l_e = \text{MAT\_Ex}[i][j][k]$ and $l_h = \text{MAT\_Hy}[i][j][k]$. Arrays MAT_Ex and MAT_Hy specify the material type associated to the relevant quantity, whereas indices i, j, k prescribe the distinct cell occupied by this medium. Hence, all parameters of (2.4)–(2.9) are turned into 1-D arrays and if the domain involves L media, then each of the ep[.], eqy[.], eqz[.], hp[.], hpx[.], hqz[.] arrays will contain only L elements which, via the l_e or l_h pointers, lead to the analogous formulae. Observe that spatial staggering is also applicable to MAT_E and MAT_H, thus imposing the staircase perspective in the construction of an EMC model.

Due to its explicit nature the FDTD algorithm is a conditionally stable scheme with the stability limit of

$$\upsilon\Delta t \leq \left[\frac{1}{(\Delta x)^2} + \frac{1}{(\Delta y)^2} + \frac{1}{(\Delta z)^2} \right]^{-1/2}, \hspace{2cm}\text{(2.13)}$$

for υ the propagation velocity in the computational lattice.

To attain a better performance in the situation of smoothly curved geometries, the technique can be alternatively derived by means of the integral—instead of the differential—version of Ampère's and Faraday's laws on electrically small orthogonal contours. Figure 2.1 shows the mutually intersecting contours as a set of chain-like coupling paths that link all field variations. This evidence, along with a careful contour deformation, can capture many shape oddities, offering valuable interpretations to difficult wave phenomena.

Selecting the time-stepping of E_y and inspecting Figure 2.1a, Ampère's law yields

$$\varepsilon \Delta x \Delta z \left[\frac{E_y|_{i,j,k}^{n+1} - E_y|_{i,j,k}^{n}}{\Delta t} \right] + \sigma \Delta x \Delta z \left[\frac{E_y|_{i,j,k}^{n+1} + E_y|_{i,j,k}^{n}}{2} \right],$$

$$= \left[H_x|_{i,j,k+1/2}^{n+1/2} - H_x|_{i,j,k-1/2}^{n+1/2} \right] \Delta x - \left[H_z|_{i+1/2,j,k}^{n+1/2} - H_z|_{i-1/2,j,k}^{n+1/2} \right] \Delta z \qquad (2.14)$$

under the constraint that E_y is invariable and equal to its average value over the surface of the cell. To complete the discretization, Faraday's law for H_x, as presented in Figure 2.3b, gives

$$\mu \Delta y \Delta z \left[\frac{H_x|_{i,j,k}^{n+1/2} - H_x|_{i,j,k}^{n-1/2}}{\Delta t} \right] + \rho' \Delta y \Delta z \left[\frac{H_x|_{i,j,k}^{n+1/2} + H_y|_{i,j,k}^{n-1/2}}{2} \right]$$

$$= \left[E_y|_{i,j,k+1/2}^{n} - E_y|_{i,j,k+1/2}^{n} \right] \Delta y - \left[E_z|_{i,j+1/2,k}^{n} - E_z|_{i,j-1/2,k}^{n} \right] \Delta z. \qquad (2.15)$$

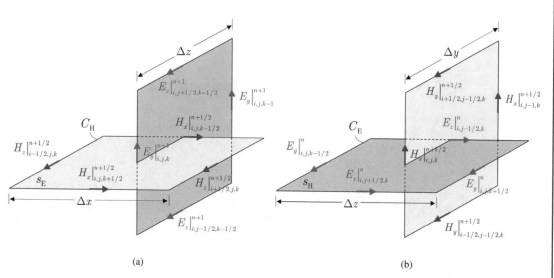

(a)   (b)

FIGURE 2.1: Graphical depiction of the contour-based FDTD formulation. (a) Ampère's law for E_y and (b) Faraday's law for H_x.

A useful hint is that the option of spatial/temporal increments should always be in compromise with the specifications of the EMC problem. Accordingly, a successful Δt is the one that guarantees the suitable grid resolution for the incorporation of every geometric detail. The essential limitation is that the cell size must be less than the smallest wavelength. So, a regular rule-of-thumb is 10 cells per wavelength λ, i.e., the side of each cell should be $\lambda/10$ or less than the wavelength of the highest frequency.

2.2.2   The Transmission-Line Matrix/Modeling Technique

The TLM method has been proven a versatile tool for rigorous EMC computations. In this algorithm, the continuous space is divided into an amount of elementary cells, designated as nodes, that form the basic mesh [4–6]. Electromagnetic vectors are then represented by an expansion in subdomain base functions, which, herein, are triangular pulses in time and a product of 2-D triangular/rectangular pulses in space. In particular, the expansions of **E** field quantities are shifted by half an increment in space and time relatively to their **H** counterparts. Starting from Maxwell's time-dependent equations, expressed as

$$\nabla \times \mathbf{H} = \frac{1}{c\eta_0}\frac{\partial \mathbf{E}}{\partial t}, \qquad \nabla \times \mathbf{E} = -\frac{\eta_0}{c}\frac{\partial \mathbf{H}}{\partial t}, \qquad (2.16)$$

with $\eta_0 = \sqrt{\mu_0/\varepsilon_0}$ and $c = 1/\sqrt{\mu_0\varepsilon_0}$, electric and magnetic field components are expanded in

$$E_x(x, y, z, t) = \sum_{i,j,k,\,n=-\infty}^{+\infty} E_x|_{i+1/2,\,j,\,k}^{n} p_{i+1/2}(x)q_j(y)q_k(z)q_n(t), \qquad (2.17)$$

$$E_y(x, y, z, t) = \sum_{i,j,k,\,n=-\infty}^{+\infty} E_y|_{i,\,j+1/2,\,k}^{n} q_i(x)p_{j+1/2}(y)q_k(z)q_n(t), \qquad (2.18)$$

$$E_z(x, y, z, t) = \sum_{i,j,k,\,n=-\infty}^{+\infty} E_z|_{i,\,j,\,k+1/2}^{n} q_i(x)q_j(y)p_{k+1/2}(z)q_n(t), \qquad (2.19)$$

$$H_x(x, y, z, t) = \sum_{i,j,k,\,n=-\infty}^{+\infty} H_x|_{i,\,j+1/2,\,k+1/2}^{n+1/2} p_i(x)q_{j+1/2}(y)q_{k+1/2}(z)q_{n+1/2}(t), \qquad (2.20)$$

$$H_y(x, y, z, t) = \sum_{i,j,k,\,n=-\infty}^{+\infty} H_y|_{i+1/2,\,j,\,k+1/2}^{n+1/2} q_{i+1/2}(x)p_j(y)q_{k+1/2}(z)q_{n+1/2}(t), \qquad (2.21)$$

$$H_z(x, y, z, t) = \sum_{i, j, k, n=-\infty}^{+\infty} H_z|_{i+1/2, j+1/2, k}^{n+1/2} q_{i+1/2}(x) q_{j+1/2}(y) p_k(z) q_{n+1/2}(t), \qquad (2.22)$$

where basis functions $p_m(u)$, with $m = i, j, k$ and $u = x, y, z$, are depicted by

$$p_m(u) = P\left(\frac{u}{\Delta u} - m\right) \text{ in which } P(u) = \begin{cases} 1 & \text{for } |u| < 1/2 \\ 1/2 & \text{for } |u| = 1/2, \\ 0 & \text{for } |u| > 1/2 \end{cases} \qquad (2.23)$$

is the rectangular pulse and basis functions $q_m(u)$, with $u = x, y, z, t$, by

$$q_m(u) = Q\left(\frac{u}{\Delta u} - m\right) \text{ in which } Q(u) = \begin{cases} 1 - |u| & \text{for } |u| < 1, \\ 0 & \text{for } |u| \geq 1 \end{cases} \qquad (2.24)$$

is the triangle pulse. Notice that $p_m(u)$ and $q_m(u)$ achieve the appropriate piecewise linear and step approximations of the exact solution of (2.16) along u-axis, respectively. The next step involves the sampling of Maxwell's equations at the cell boundaries through delta functions in space and time. For example, the sampling of the $\partial E_z / \partial t$ derivative is performed by

$$\text{Sampling}\left\{\frac{\partial E_z}{\partial t}\right\} \equiv \iiiint \frac{\partial E_z}{\partial t} \delta(x - i\Delta x)\delta(y - j\Delta y)\delta(z - k\Delta z)\delta(t - n\Delta t) dx dy dz dt, \qquad (2.25)$$

with the following integrals taken into consideration

$$\int_{-\infty}^{+\infty} \delta[u - (l + \tau)\Delta u] p_m(u) du = \delta_{l,m},$$

$$\int_{-\infty}^{+\infty} \delta(u - l\Delta u) q_m(u) du = \delta_{l,m},$$

$$\int_{-\infty}^{+\infty} \delta[u - (l + 1/4)\Delta u] q_m(u) du = \frac{1}{4}(3\delta_{l,m} + \delta_{l+1,m}),$$

$$\int_{-\infty}^{+\infty} \delta[u - (l + 3/4)\Delta u] q_m(u) du = \frac{1}{4}(\delta_{l,m} + 3\delta_{l+1,m}),$$

$$\int\limits_{-\infty}^{+\infty} \delta\left[u - (l + 1/2 + \tau)\Delta u\right] \frac{\partial q_m(u)}{\partial u} du = \delta_{l+1, m} - \delta_{l, m},$$

where $\delta_{l, m}$ is the Kronecker delta; $l = i, j, k$; $m = i, j, k$ (selected toward a specific direction of the lattice axes); $u = x, y, z, t$; and $\tau = 0, \pm 1/4$. This procedure, also applied to $\partial H_x/\partial y$ and $\partial H_y/\partial x$, leads to *eight* discrete expressions which, if added, give

$$E_z\big|_{i+1, j, k+1/2}^{n+1} + E_z\big|_{i-1, j, k+1/2}^{n+1} + E_z\big|_{i, j+1, k+1/2}^{n+1} + E_z\big|_{i, j-1, k+1/2}^{n+1} + 12\, E_z\big|_{i, j, k+1/2}^{n+1}$$

$$- E_z\big|_{i+1, j, k+1/2}^{n} - E_z\big|_{i-1, j, k+1/2}^{n} - E_z\big|_{i, j+1, k+1/2}^{n} - E_z\big|_{i, j-1, k+1/2}^{n} - 12\, E_z\big|_{i, j, k+1/2}^{n}$$

$$= \kappa\eta \left(H_y\big|_{i+1/2, j, k+1/2}^{n+3/2} + 6\, H_y\big|_{i+1/2, j, k+1/2}^{n+1/2} + H_y\big|_{i+1/2, j, k+1/2}^{n-1/2} \right)$$

$$- \kappa\eta \left(H_y\big|_{i-1/2, j, k+1/2}^{n+3/2} + 6\, H_y\big|_{i-1/2, j, k+1/2}^{n+1/2} + H_y\big|_{i-1/2, j, k+1/2}^{n-1/2} \right)$$

$$- \kappa\eta \left(H_x\big|_{i, j+1/2, k+1/2}^{n+3/2} + 6\, H_x\big|_{i, j+1/2, k+1/2}^{n+1/2} + H_x\big|_{i, j+1/2, k+1/2}^{n-1/2} \right)$$

$$+ \kappa\eta \left(H_x\big|_{i, j-1/2, k+1/2}^{n+3/2} + 6\, H_x\big|_{i, j-1/2, k+1/2}^{n+1/2} + H_x\big|_{i, j-1/2, k+1/2}^{n-1/2} \right) \tag{2.26}$$

For $\kappa = 2\eta_0 c\Delta t/\eta\Delta$, Δ is the spatial increment of a cubic mesh and η is an arbitrary impedance. In an effort to properly relate wave amplitudes to tangential electric and magnetic field quantities at the boundary surfaces, a cell boundary approach is developed. According to its aspects, every component is written as the result of its series expansion at the boundary of the cell. Consequently, for the E_z terms of (2.26), it holds that

$$E_z\big|_{i\pm1/4, j, k+1/2}^{n} = \frac{1}{4}\left(E_z\big|_{i\pm1, j, k+1/2}^{n} + 3\, E_z\big|_{i, j, k+1/2}^{n} \right),$$

$$E_z\big|_{i, j\pm1/4, k+1/2}^{n} = \frac{1}{4}\left(E_z\big|_{i, j\pm1, k+1/2}^{n} + 3\, E_z\big|_{i, j, k+1/2}^{n} \right), \tag{2.27}$$

whereas for the H_x and H_y ones

$$H_x\big|_{i, j+1/4, k+1/2}^{n+1/2} = \frac{1}{4}\left(3\, H_x\big|_{i, j+1/2, k+1/2}^{n+1/2} + H_x\big|_{i, j-1/2, k+1/2}^{n+1/2} \right), \tag{2.28}$$

$$H_x\big|_{i, j-1/4, k+1/2}^{n+1/2} = \frac{1}{4}\left(H_x\big|_{i, j+1/2, k+1/2}^{n+1/2} + 3\, H_x\big|_{i, j-1/2, k+1/2}^{n+1/2} \right), \tag{2.29}$$

$$H_y\big|_{i+1/4, j, k+1/2}^{n+1/2} = \frac{1}{4}\left(3\, H_x\big|_{i+1/2, j, k+1/2}^{n+1/2} + H_x\big|_{i-1/2, j, k+1/2}^{n+1/2} \right), \tag{2.30}$$

$$H_y\Big|_{i-1/4,\,j,\,k+1/2}^{n+1/2} = \frac{1}{4}\left(H_x\Big|_{i+1/2,\,j,\,k+1/2}^{n+1/2} + 3\,H_x\Big|_{i-1/2,\,j,\,k+1/2}^{n+1/2}\right) \tag{2.31}$$

To this end, analogous values are defined for the temporal variation in (2.26), as

$$E_z\Big|_{i,\,j,\,k}^{n+1/4} = \frac{1}{4}\left(E_z\Big|_{i,\,j,\,k}^{n+1} + 3\,E_z\Big|_{i,\,j,\,k}^{n}\right), \quad E_z\Big|_{i,\,j,\,k}^{n+3/4} = \frac{1}{4}\left(3\,E_z\Big|_{i,\,j,\,k}^{n+1} + E_z\Big|_{i,\,j,\,k}^{n}\right), \tag{2.32}$$

$$H_{x,y}\Big|_{i,\,j,\,k}^{n+1/4} = \frac{1}{4}\left(3\,H_{x,y}\Big|_{i,\,j,\,k}^{n+1/2} + H_{x,y}\Big|_{i,\,j,\,k}^{n-1/2}\right), \; H_{x,y}\Big|_{i,\,j,\,k}^{n+3/4} = \frac{1}{4}\left(H_{x,y}\Big|_{i,\,j,\,k}^{n+3/2} + 3\,H_{x,y}\Big|_{i,\,j,\,k}^{n+1/2}\right), \tag{2.33}$$

Substitution of (2.32), (2.33) in (2.16) gives

$$\begin{aligned}
E_z\Big|_{i+1/4,\,j,\,k+1/2}^{n+3/4} &+ E_z\Big|_{i-1/4,\,j,\,k+1/2}^{n+3/4} + E_z\Big|_{i,\,j+1/4,\,k+1/2}^{n+3/4} + E_z\Big|_{i,\,j-1/4,\,k+1/2}^{n+3/4} \\
&- E_z\Big|_{i+1/4,\,j,\,k+1/2}^{n+1/4} - E_z\Big|_{i-1/4,\,j,\,k+1/2}^{n+1/4} - E_z\Big|_{i,\,j+1/4,\,k+1/2}^{n+1/4} - E_z\Big|_{i,\,j-1/4,\,k+1/2}^{n+1/4} \\
&= \kappa\eta\left(H_y\Big|_{i+1/4,\,j,\,k+1/2}^{n+3/4} + H_y\Big|_{i+1/2,\,j,\,k+1/2}^{n+1/4} - H_y\Big|_{i-1/4,\,j,\,k+1/2}^{n+3/4} - H_y\Big|_{i-1/2,\,j,\,k+1/2}^{n+1/4}\right) \\
&- \kappa\eta\left(H_x\Big|_{i,\,j+1/4,\,k+1/2}^{n+3/4} + H_x\Big|_{i,\,j+1/4,\,k+1/2}^{n+1/4} - H_x\Big|_{i,\,j-1/4,\,k+1/2}^{n+3/4} - H_x\Big|_{i,\,j-1/4,\,k+1/2}^{n+1/4}\right)
\end{aligned} \tag{2.34}$$

The major TLM advantage is its efficiency to handle involved devices. Exhibiting a considerable universality, it includes the properties of $\mathbf{E}$, $\mathbf{H}$ vectors and their interactions with all boundaries and material interfaces. Thus, it is not mandatory to reformulate the problem under study for every minor modification of its parameters, because no loss of convergence, instabilities, or oscillatory wave fronts are created.

2.2.3   The Pseudospectral Time-Domain Method

When realistic EMC structures with prolonged simulation times are to be discretized, it is generally mandatory to raise mesh resolution for the control of the cumulative reflection errors. To overcome this tough requisite, without sacrificing other benefits of time-domain analysis, the PSTD technique seems to be a promising solution [7, 8]. Rather than the customary finite-difference approximators, this method uses either trigonometric functions or Chebyshev polynomials, through a fast Fourier transform (FFT), for the representation of spatial derivatives. Because its advanced spectral accuracy, the PSTD technique needs only two cells per wavelength consistent with the Nyquist sampling theorem and, therefore, can handle more difficult problems. One critical limitation, however, is the spatial periodicity stipulation imposed by the use of the FFT.

Let us first write Maxwell's equations in terms of a general coordinate-stretching system, which launches

$$\xi_u = b_u + j\frac{\omega_u}{\omega} \quad \text{for } u = x, y, z, \qquad (2.35)$$

and b_u an independent scaling factor that speeds up the suppression of artificial wave fronts, like evanescent ones, in lossy materials. Replacing the curl operator in Ampère's and Faraday's law with

$$\nabla_\xi \times [.] \equiv \sum_{u=x,\,y,\,z} \frac{1}{\xi_u}\frac{\partial}{\partial u}[.]\hat{\mathbf{u}},$$

it can be promptly acquired that

$$\varepsilon b_u \frac{\partial \mathbf{E}}{\partial t} + (b_u\sigma + \omega_u\varepsilon)\mathbf{E} = \frac{\partial}{\partial t}(\hat{\mathbf{u}} \times \mathbf{H}) - \omega_u\sigma \int_{-\infty}^{t} \mathbf{E}\mathrm{d}t, \qquad (2.36)$$

$$\mu\left(b_u\frac{\partial \mathbf{H}}{\partial t} + \omega_u\mathbf{H}\right) = -\frac{\partial}{\partial u}(\hat{\mathbf{u}} \times \mathbf{E}) - \mathbf{M}_c \qquad (2.37)$$

In contrast to the FDTD method, which forms a staggered grid, the PSTD approach uses a lattice where all $\mathbf{E}$ and $\mathbf{H}$ components are positioned at the cell centers. Such a mesh offers the possibility of determining media properties and mitigating field singularities in curvilinear coordinates. If the computational domain is divided into $M_x \times M_y \times M_z$ cells, then spatial derivatives are approximated by

$$\frac{\partial f(\mathbf{r})}{\partial u} = \frac{2\pi}{M_u\Delta u}\mathcal{F}_u^{-1}\{jk_u\mathcal{F}_u\{f(\mathbf{r})\}\} \quad \text{with} \quad \mathcal{F}_u\{f(\mathbf{r})\} \equiv \int_{-\infty}^{+\infty} f(\mathbf{r})\mathrm{e}^{-jk_u u}\mathrm{d}u \quad (2.38)$$

the Fourier transform of any function $f(\mathbf{r})$, $\mathcal{F}_u^{-1}\{.\}$ is the respective inverse transform, Δu is the spatial increment along u, and k_u is the index of FFT that takes places for all u along a straight line cut through the plane of the other two coordinates. It is important to stress that the derivative in (2.38) is exact as long as $\Delta u \leq \lambda/2$ (λ is the wavelength). So, even for two cells per wavelength, PSTD schemes do not cause any dispersion errors. For instance, considering trigonometric functions, spatial derivatives are given by

$$\left.\frac{\partial f(\mathbf{r})}{\partial u}\right|_{s\Delta u} = \frac{1}{M_u\Delta u}\sum_{m=-M_u/2}^{M_u/2-1} jk_{u,m}\tilde{f}(m)\mathrm{e}^{jk_{u,m}s\Delta u}, \qquad (2.39)$$

where $k_{u,m} = 2\pi m / (M_u \Delta_u)$ and $\tilde{f}(m)$ is the Fourier series

$$\tilde{f}(m) = \Delta u \sum_{s=0}^{M_u - 1} f(s\Delta u)e^{-jk_{u,m}s\Delta u}$$

In this way, equations (2.36) and (2.37) become

$$\varepsilon b_u \frac{\partial \mathbf{E}}{\partial t} + (b_u \sigma + \omega_u \varepsilon)\mathbf{E} = -\hat{\mathbf{u}} \times \mathcal{F}_u^{-1}\{jk_{u}\mathcal{F}_u\{\mathbf{H}\}\} - \omega_u \sigma \int_{-\infty}^{t} E dt, \qquad (2.40)$$

$$\mu \left(b_u \frac{\partial \mathbf{H}}{\partial t} + \omega_u \mathbf{H} \right) = \hat{\mathbf{u}} \times \mathcal{F}_u^{-1}\{jk_{u}\mathcal{F}_u\{\mathbf{E}\}\} - \mathbf{M}_c \qquad (2.41)$$

To complete the update, $\mathbf{E}$ components are defined at $t = n\Delta t$, whereas $\mathbf{H}$ ones at $t = (n + 1/2)\Delta t$. For the rest of the terms in (2.40) and (2.41), not available at the required time steps, an averaging scheme, which does not degrade discretization quality, is developed. The prior formulation entails only 1-D FFTs, thus permitting fast computations with an order of $O(M_u \log_2 M_u)$ operations. Concerning the stability of the PSTD algorithm, the necessary criterion in a lossless, homogeneous domain is

$$\upsilon \Delta t \leq \frac{2\Delta u}{\pi \sqrt{\dim}}, \qquad (2.42)$$

with dim as the dimensionality of the problem. Evidently, (2.42) is a bit stricter than the FDTD Courant limit because of the $2/\pi$ term. Nonetheless, such a restriction does not create any difficulties, because for electrically large structures, Δt is dominated by accuracy aspects rather than the numerical stability condition itself.

2.3   CALCULATION OF TRANSMISSION-LINE AND S-PARAMETERS

A very frequent request during the investigation of a PCB structure is the computation of transmission line and (scattering) S-parameters in a broadband frequency spectrum. Concentrating on the former and presuming systems with signal propagation in the TEM mode, the subsequent integral relations

$$V(x_i, t) = \int_{C_V} \mathbf{E}(x_i, t) \cdot d\mathbf{l}, \quad I(x_i, t) = \int_{C_I} \mathbf{H}(x_i, t) \cdot d\mathbf{l}, \qquad (2.43)$$

link electric and magnetic intensity vectors to the voltage and current circuit quantities. In (2.43), C_V is the contour that spans—in a perpendicular way—from a prefixed voltage reference point to the PCB conductor at the neighboring position x_s. Also, C_I is the contour that totally surrounds the PCB conductor in the transverse plane at the smallest possible distance from its surface. So, the line characteristic impedance is derived by

$$Z_0(x_i, \omega) = \frac{\mathcal{F}_x\{V_{\text{inc}}(x_i, t)\}}{\mathcal{F}_x\{I_{\text{inc}}(x_i, t)\}}, \tag{2.44}$$

where $V_{\text{inc}}(x_i, t)$ and $I_{\text{inc}}(x_i, t)$ are the incident voltage and current obtained at x_i. For an impulsive excitation, the general expression of the reflection coefficient anywhere in the transmission line model is given by

$$\Gamma(x_i, \omega) = \frac{\mathcal{F}_x\{V_{\text{ref}}(x_i, t)\}}{\mathcal{F}_x\{V_{\text{inc}}(x_i, t)\}}, \tag{2.45}$$

with $V_{\text{ref}}(x_i, t)$ the reflected voltage at x_s. On the other hand, the complex S_{mm} parameters in a multiport device are evaluated in terms of

$$S_{lm}(x_l, x_m, \omega) = \frac{\dot{V}_l(x_l, \omega)}{\dot{V}_m(x_m, \omega)} \left[\frac{\bar{Z}_{0,m}(\omega)}{\bar{Z}_{0,l}(\omega)}\right]^{1/2}, \tag{2.46}$$

in which $\dot{V}_s(x_s, \omega)$, for $s = l$, m, is the phasor voltage at port s and plane x_s, whereas $Z_{0,s}(\omega)$ is the characteristic impedance of the line connected to the respective port. Actually, contrary to the phase, the magnitude of S parameters is of real interest from an engineering perspective, because it is independent of the observation plane on condition that the feeding lines have infinite lengths or they are matched at their far ends.

2.4   LOW-PASS SPATIAL FILTERING SCHEMES

A critical factor that may have a destabilizing effect on time-domain simulations is the case of highly nonuniform grids—often encountered in EMC computational models—created by the arbitrary alignment of material interfaces. One viable approach can be pursued in the annihilation of spurious frequencies that are likely to arise in the vicinity of the aforementioned boundaries. For this purpose, a postprocessing filtering algorithm, applied to the cell average $\bar{G}_q$, is used. According to its basic premises, one starts from

$$\bar{G}_q = \frac{1}{\Delta q} \int_{q_1}^{q_2} G dq, \tag{2.47}$$

with the limits of integration $[q_1, q_2]$ being the dimensions $[u_1, u_2]$, $[v_1, v_2]$, $[w_1, w_2]$ of the cell edges, defined in the general coordinate system (u, v, w), Δq the spatial increment, and G any component of vectors $\mathbf{E}, \mathbf{H}$. In the case of multispace problems, (2.47) is used sequentially along the directions u, v, w in the lattice.

Introducing an additional degree of freedom, Ξ (to be considered as the approximation order), for the control of filtering, the expression in the interior of the domain receives the following form

$$\bar{G}'_{s-1} + \bar{G}'_s + \bar{G}'_{s+1} = \frac{1}{2} \sum_{\xi=0}^{\Xi} a_\xi \left(\bar{G}_{s+\xi} + \bar{G}_{s-\xi} \right) \quad s = i, j, k \tag{2.48}$$

where the primed quantity represents the filtered field value. The frequency response of (2.48) is

$$FS(k) = \frac{\sum\limits_{\xi}^{\Xi} a_\xi \cos(\xi k)}{1 + 2 \cos(k)}, \tag{2.49}$$

which has $\Xi + 2$ unknowns. Owing to the symmetric profile of (2.48), FS(k) is real and thus the filter acts only on the magnitude of every field quantity. To suppress the highest frequency mode, it is imposed that $FS(k) = 0$ in order to obtain $\Xi + 1$ equations by matching the appropriate Taylor series expanded about s. A typical value is $\Xi = 4$ with $a_0 = 1.2875874$, $a_1 = -1.9237896$, $a_2 = 2.5751793$, and $a_3 = 0.3468214$, whereas a complete set of filtering coefficients for various approximation orders Ξ is given in Table 2.1. Note that the above filter does not seriously affect the operational cost of the simulation.

TABLE 2.1: Filtering coefficients of diverse approximation orders

ORDER Ξ	a_0	a_1	a_2	a_3	a_4	a_5	a_6
2	0.5413219	−3.6391002	0	0	0	0	0
3	−2.4987732	1.2469018	−0.9264315	0	0	0	0
4	1.2675874	−1.9237896	2.5751793	0.3468214	0	0	0
5	−0.4137282	4.0248721	−0.5137282	2.8104217	−1.9423155	0	0
6	3.4900416	−2.8760425	1.7976014	−4.4390781	5.2145871	2.7629021	0
7	−0.8247458	1.0481972	−6.1017825	3.5121564	−0.7536906	7.0136185	−5.1382374

2.5   DC-POWER BUS CONFIGURATIONS

The first category of PCB applications are the DC-power bus arrangements that constitute a very important part of most modern EMC structures, because they are responsible for the correct energy flow in the device [9–15]. Let us investigate the 3-D PCB problem in Figure 2.2 comprising an arbitrary number of integrated circuits N. These circuits, mounted on the first layer, consist of the same material that has the following properties: $\varepsilon_A = 2.5\varepsilon_0$, $\mu_A = 1.76\mu_0$, $\sigma_A = 0.22$ S/m, whereas those of the substrate are $\varepsilon_B = 4.7\varepsilon_0$, $\mu_B = \mu_0$, $\sigma_B = 0.03$ S/m. Moreover, the decoupling capacitances of $0.01\mu F$ are used to supply the return pass of high frequency energy, whereas for the excitation a Gaussian pulse is used. The overall dimensions are $b = 16.39$ mm, $b_1 = 18.65$ mm, $b_2 = 2.5$ mm, and for the simulation the PSTD and the FDTD method have been selected with $\Delta x = \Delta y = \Delta z = 1.5$ mm and $\Delta t = 25.32$ ps.

Figure 2.3a and 2.3b demonstrates the variation of S_{11} and S_{21} parameters as a function of frequency for different time-domain realizations. The reference solution has been retrieved from Refs. [10, 15]. As can be promptly deduced, both schemes exhibit a very good accuracy, without any discrepancies in the frequency spectrum.

It should be mentioned, though, that the PSTD algorithm is proven to be more robust from an implementation viewpoint due to the efficient handling of spatial derivatives and the enhanced resolutions it can offer for the same amount of cells. This is a very instructive deduction especially in the analysis of large-scale EMC problems where traditional techniques lack to provide adequate solutions. To this end, Figure 2.4a and 2.4b depicts the bus' input resistance in terms of dimension b and the number of circuits N, respectively. Evidently, the variations are less prominent when b is greater and total performance degrades as N increases.

Next, the hemiellipsoidal cavity-backed dielectric resonator antenna fed by a PCB-power bus (Figure 2.5) is examined. The cover of the antenna has a relative dielectric permittivity of $\varepsilon_{rc} = 8.6$

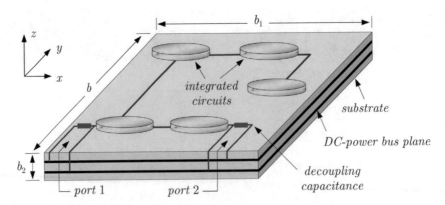

FIGURE 2.2: A power bus printed circuit board with all of its integrated circuits consisting of the same lossy material.

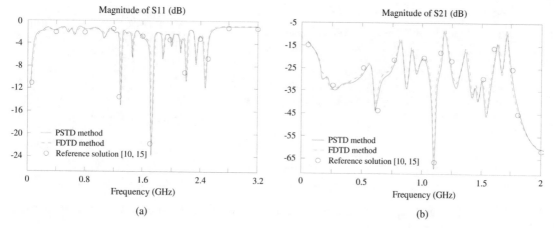

FIGURE 2.3: Magnitude of (a) S11 parameter and (b) S21 parameter for the PCB-power bus device.

and a patch attached on its surface at $\varphi = 45°$ in order to attain optimal beam steering. In additional, the structure has: $l_A = 48.5$ mm, $l_B = 54$ mm, $r = 6.8$ mm, $s = 1.08$ mm, and $t = 0.96$ mm. Figure 2.6a gives the return loss of the antenna and Figure 2.6b its axial ratio and gain for a patch of $d_1 = 20.45$ mm, $d_2 = 10$ mm.

Simulations are performed via the FDTD (grid of $230 \times 242 \times 84$ cells) and the finite-element time-domain (FETD; mesh of 258,932 elements) method with and without the aid of the low-pass spatial filtering process discussed in Section 2.4. Concerning the open boundaries

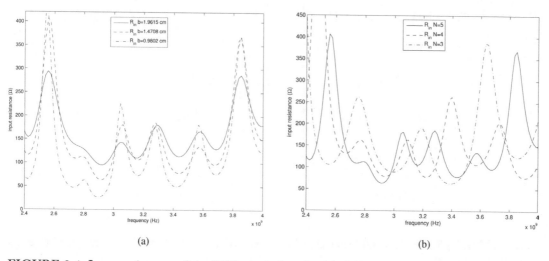

FIGURE 2.4: Input resistance of the PCB-power bus for (a) different b and (b) diverse number of circuits N.

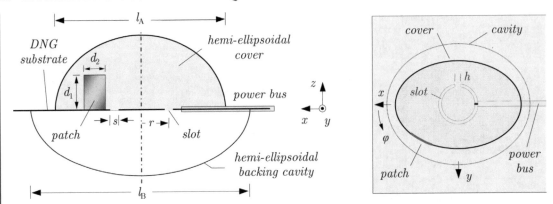

FIGURE 2.5: Side and top views of a hemiellipsoidal cavity-backed antenna with one patch and a PCB-power bus at its feeding port.

of the domain, these are effectively terminated by a six-cell perfectly matched layer (PML) absorber tailored to cooperate with both schemes. The improvement in the accuracy of the outcomes is obvious, despite the challenging shape and material distributions of the problem. Similar conclusions are easily extracted from Figure 2.7, which presents the radiation patterns of the particular antenna for the frequency of 3.9 GHz at different planes.

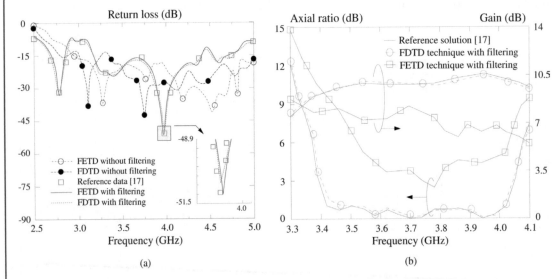

FIGURE 2.6: (a) Return loss and (b) axial ratio and gain of the hemiellipsoidal antenna fed by means of the PCB-power bus.

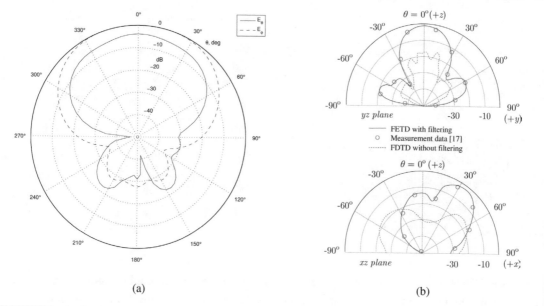

(a)   (b)

FIGURE 2.7: Radiation patterns of the hemiellipsoidal helical antenna at 3.9 GHz. (a) *xy* plane and (b) *yz* and *xz* planes.

2.6   DECOUPLING DEVICES AND MULTILAYERED SUBSTRATES

Another noteworthy class of modern EMC components is the decoupling devices, usually fabricated on multilayered substrates [16–22]. Based on their notable wideband competences, these structures are found to be fairly useful in the design of ultrabandwidth microwave elements, junctions, and couplers or even shielding components. Assume the cylindrically shielded multiconductor transmission line of Figure 2.8a.

The analysis of this complicated problem is performed via the TLM and the FETD algorithms, whereas for the sake of comparison the FDTD method and the technique of [21] are also used. The basic dimensions are $L = 5.5$ mm, $g = 2.28$ mm, $d_1 = 1.43$ mm, $d_2 = 1.12$ mm, $d_3 = 1.43$ mm, with the first substrate ($\varepsilon_1 = 2.35\varepsilon_0$) being $l_1 = 2.46$ mm thick and the second ($\varepsilon_2 = 1.78\varepsilon_0$) having a depth of $l_2 = 2.15$ mm. Moreover, along the z-direction the device is truncated by an eight-layer PML. Under these circumstances, Figure 2.8b presents the insertion loss of the transmission line for diverse configurations. Inspecting the outcomes, one can discern the promising agreement of the TLM and the FETD plots with the method of [21] regarded as our reference. On the contrary, the conventional FDTD algorithm is proven to be less accurate, an issue accredited to the inability of its "staircase" discretization to provide adequate models for the cylindrical

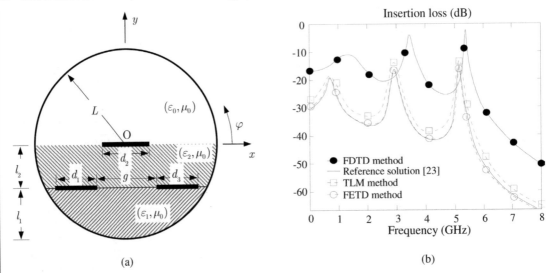

(a) (b)

FIGURE 2.8: (a) Cross section of a cylindrically shielded dual-plane decoupling device with three microstrip lines and (b) computation of its insertion loss through different time-domain methods.

TABLE 2.2: First resonance frequency for the cylindrically shielded dual-layered decoupling device

REFERENCE (GHz) [21]	METHOD	RESONANCE (GHz)	ERROR (%)	MAXIMUM GLOBAL ERROR
Case A 5.228	TLM	5.131	1.863	2.257×10^{-2}
	FETD	5.138	1.715	2.485×10^{-2}
	FDTD	4.840	7.421	5.314×10^{-1}
Case B 7.439	TLM	7.293	1.953	3.638×10^{-2}
	FETD	7.276	2.185	3.792×10^{-2}
	FDTD	6.701	9.926	9.643×10^{-1}
Case C 9.675	TLM	9.447	2.352	5.871×10^{-2}
	FETD	9.437	2.458	5.913×10^{-2}
	FDTD	8.644	10.653	1.174

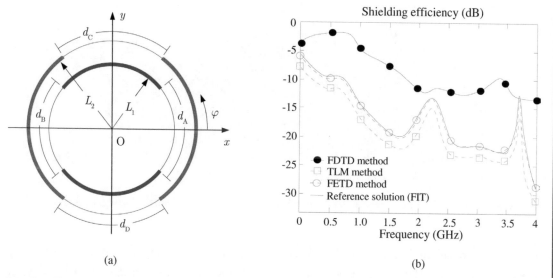

(a)   (b)

FIGURE 2.9: (a) A two-layered cylindrical decoupling cavity with four co-centric microstrip elements and (b) its shielding efficiency.

shield. Furthermore, Table 2.2 summarizes the first resonance frequency of three different cases with $L = 8.5$ mm and $g = 1.6$ mm (case A: $d_1 = 1.75$ mm, $d_2 = 1.26$ mm, $d_3 = 1.75$ mm; case B: $d_1 = 2.38$ mm, $d_2 = 1.62$ mm, $d_3 = 2.38$ mm; case C: $d_1 = 3.22$ mm, $d_2 = 1.84$ mm, $d_3 = 3.22$ mm) for the same substrate characteristics along with the maximum global error of the preceding numerical methods and compares the results with those obtained via FIT (reference solution). Again, the TLM and the FETD formulations lead to the most precise solutions for the immunity behavior with very logical global-error levels.

Subsequently, let us consider the cylindrical decoupling cavity of Figure 2.9a. The lengths d_A, d_B and d_C, d_D correspond to the angles of 75° and 80° with radii $L_1 = 10$ mm and $L_2 = 12$ mm, respectively. The arrangement is illuminated by a plane wave incorporating a variety of evanescent modes and terminated by a six-cell PML. Once more, the outcomes of the TLM, FETD, and FDTD (lattice of $84 \times 84 \times 250$ cells) techniques are compared to an FIT solution based on an adequately fine grid. Figure 2.9b shows the shielding efficiency, where the values, obtained by the first two algorithms, are found to be the most rigorous.

2.7 SWITCHING INTERCONNECTS AND INTEGRATED VIAS

The presence of metallization surfaces in the form of planar power planes, ground layers, or vertical vias and ports constitutes a powerful means in the arsenal of microwave engineers for the construction of highly competent devices in many contemporary EMC designs [23–31]. Owing to their

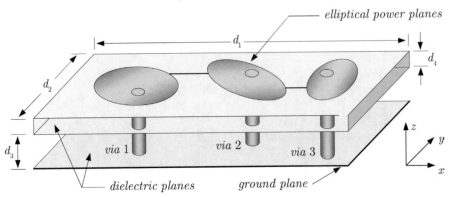

FIGURE 2.10: A PCB arrangement with switching interconnects and vertical integrated vias.

miniaturized details and thin layers, these PCBs require a meticulous investigation, before entering the factory line, in order to subdue all undesired oscillatory wave mechanisms. Actually, it is the impact of these artificialities that creates rapid field changes near the vias or the discontinuities and therefore spoils the performance of the entire structure.

Our analysis concentrates on the general PCB layout of Figure 2.10, which involves elliptical power planes and vias. It also has a properly selected set of dielectric planes with a relative

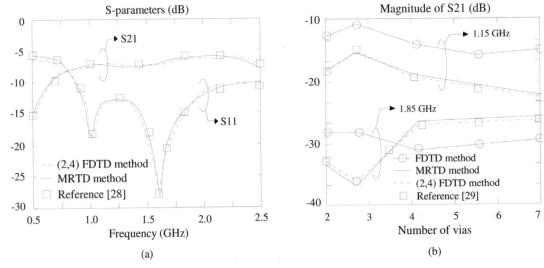

FIGURE 2.11: Magnitude of (a) S11 and S12 parameter versus frequency and (b) S21 parameter versus the number of vias of the general PCB arrangement.

dielectric permittivity of $\varepsilon_r = 3.2$ and conductivity $\sigma = 0.05$ S/m. Its dimensions are: $d_1 = 28$ cm, $d_2 = 15$ cm, $d_3 = 4.5$ cm, and $d_4 = 1.7$ cm, whereas for its simulation the higher-order (2, 4) FDTD method combined with the spatial filtering procedure and an 8-layer PML, is implemented. Moreover, the results, so derived, are compared to the ones of the MRTD method (see Chapter 3). In this manner, the domain is divided into $82 \times 86 \times 64$ cells with $\Delta x = \Delta y = \Delta z = 0.022$ cm. Figure 2.11a and 2.11b shows the variation of different S-parameters versus frequency and number of vias, respectively. As observed, higher-order outcomes exhibit a very good precision without any inconsistencies in the frequency spectrum and the most important; they offer substantial memory savings. The MRTD method, exploiting the advanced capabilities of wavelet functions, attains equivalently acceptable results that are in very good agreement with the reference solution [26].

2.8 MONOLITHIC MICROWAVE INTEGRATED CIRCUITS

The contribution of PCB structures in the technology of monolithic microwave integrated circuits (MMIC) is, nowadays, indisputable, especially in cell phones and telecommunication networks [32–41]. This is basically attributed to the ease of fabrication and the flexibility for modular use in related designs, such as antennas, microwave amplifiers, mixers, or filters, to name a few. Moreover, the impressive evolution of photolithography and etching techniques for the realization of the above components has permitted proficient extensions to coplanar waveguides and striplines. Nonetheless, as popular as the MMIC applications are, their compliance with the contemporary EMC standards should be an issue of serious concern, because an incorrect configuration can lead to frequency-spectrum confinement, gain reduction, and inadequate signal-to-noise ratios. Bearing in mind the constantly increasing use of MMIC in wideband systems, the need for concise (if not optimal) implementations has become more persistent than ever. Thus, to meet these stipulations and monitor their complicated behavior without the need of excessive manufacturing costs, these structures must be systematically simulated by means of advanced methodologies. To this aim, time-domain algorithms can play a critical role, because they offer fast and rigorous answers to many MMIC design difficulties or other practical considerations.

2.8.1 Microstrip Antennas in EMC Arrangements

This section is devoted to the design and time-domain simulation of microstrip MMIC antennas equipped with certain PCB parts, pertinent for operation in the area of wireless local area network systems [32, 33]. The first device is the patch antenna of Figure 2.12, which is based on the use of a parasitic element for its dual-band function. The specific process has been proven very profitable, because the

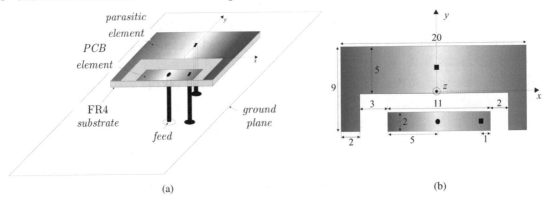

(a)   (b)

FIGURE 2.12: A dual-band MMIC patch antenna. (a) Perspective and (b) top view of the radiating-element system. All dimensions are given in mm.

presence of such an element on or near the PCB parts leads to the extension of the frequency spectrum due to the appearance of additional modes adjacent to the dominant one.

The dimensions of the antenna for its operation at 2.44 and 5.80 GHz are given in Figure 2.12b, whereas both metallic elements are printed on a 0.4-mm-thick FR4 substrate with $\varepsilon_r = 4.4$. Furthermore, the substrate is located 4.6 mm above the ground plane and the total area of the radiator is 9×20 mm². To analyze this MMIC structure from an EMC viewpoint, the FDTD technique combined with an eight-cell PML is implemented. The domain comprises $120 \times 96 \times 72$

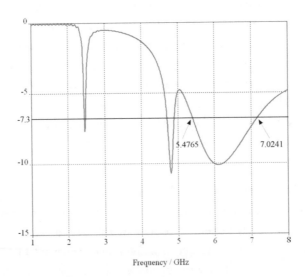

Frequency / GHz

FIGURE 2.13: Reflection coefficient of the dual-band MMIC patch antenna.

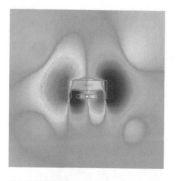 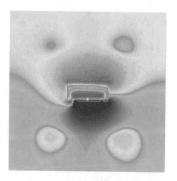 

FIGURE 2.14: Snapshots of the E_x, E_y, E_z components at the transverse to the z-axis plane that bisects the substrate for the frequency of 2.44 GHz.

cells and the temporal increment is set at 0.2164 ms. Figure 2.13 provides the reflection coefficient of the MMIC arrangement from which the two resonance frequencies at 2.44 and 5.80 GHz may be clearly discerned. Notice the confined bandwidth of the former and the almost 1.5 GHz of the latter at the level of −7.3 dB. It is stated that the extra resonance at 4.8 GHz is accredited to a higher mode of the parasitic element. Also, Figures 2.14 and 2.15 illustrate several snapshots of the three electric field intensity components at the transverse to the z-axis plane that bisects the substrate for both operating frequencies. Notice the smoothness of the plots along with the satisfactory depiction of the propagating wave fronts. Similar deductions are extracted from the E_x, E_y, E_z snapshots of Figures 2.16 and 2.17 at the $x = 0$ and $y = 0$ planes, respectively.

The second MMIC layout is the single-via meander antenna of Figure 2.18a, which is constructed by means of a helical slit on a perfectly electrically conducting plane mounted on a $\varepsilon_r = 9.2$

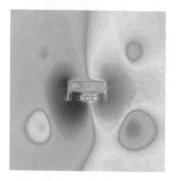 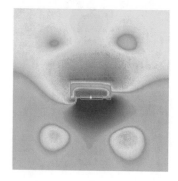 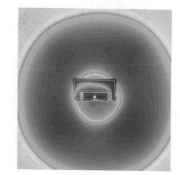

FIGURE 2.15: Snapshots of the E_x, E_y, E_z components at the transverse to the z-axis plane that bisects the substrate for the frequency of 5.80 GHz.

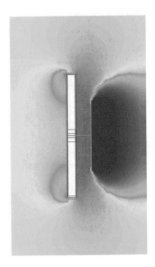

FIGURE 2.16: Snapshots of the E_x, E_y, E_z components at the $y = 0$ plane for the frequency of 2.44 GHz.

dielectric substrate with a thickness of $h = 1.905$ mm. Apart from the dimensions shown in Figure 2.18, some supplementary geometric data are: $W_1 = 0.5$ mm, $L_1 = 0.75$ mm, $S = 4$ mm, $ds = 2$ mm, and $L_2 = 2$ mm. The antenna is fed through a microstrip line terminated by a short cut at its one end. The simulation of the discretized model is performed in terms of the FIT that uses an appropriately

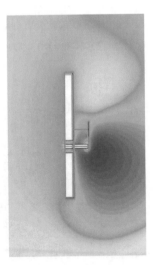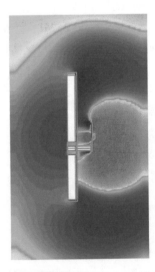

FIGURE 2.17: Snapshots of the E_x, E_y, E_z components at the $x = 0$ plane for the frequency of 5.80 GHz.

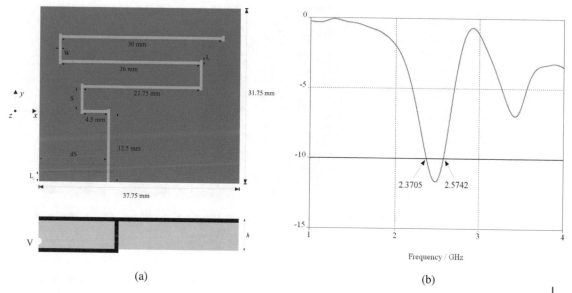

(a)   (b)

FIGURE 2.18: (a) Geometry and (b) reflection coefficient of a single-via meander MMIC antenna.

formed $110 \times 104 \times 28$ lattice. In this context, Figure 2.18b presents the reflection coefficient of the radiator and demonstrates the 200-MHz bandwidth around the frequency of 2.44 GHz at the level of -10 dB. Moreover, a variety of E_x, E_y, E_z snapshots at several planes are provided in Figures 2.19–2.21.

A more complicated antenna that can operate in a dual-band policy is next investigated [33]. It is actually a PCB radiator with two monopoles forming a T arm, as depicted in Figure 2.22a.

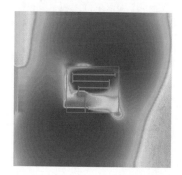
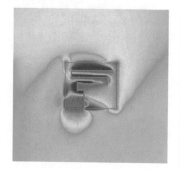

FIGURE 2.19: Snapshots of the E_x, E_y, E_z components at the transverse to the z-axis plane that bisects the substrate for the frequency of 2.44 GHz.

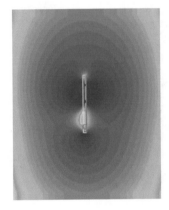
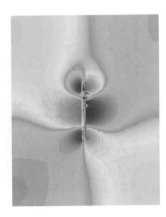

FIGURE 2.20: Snapshots of the *Ex*, *Ey*, *Ez* components at the *x* = 0 plane for the frequency of 2.44 GHz.

The largest monopole, comprising a perpendicular strip of width w and height h_1 at the center and a horizontal strip of width w_1 and length $2l_1 + w$ at the top, controls the low-frequency band. The response of the antenna at higher frequencies is adjusted by the smaller monopole that consists of a horizontal strip of width w_2 and length $l_{21} + w + l_{22}$ along with a perpendicular strip of height h_2. The choice of nonequal lengths l_{21} and l_{22} can enhance the operation bandwidth of the structure in the area of 5.25 GHz. For the excitation, a microstrip line of 50 Ω and width $w_f = 1.5$ mm is used, whereas the entire setup is printed on a 0.8-mm-thick FR4 ($\varepsilon_r = 4.4$) substrate.

The rest of the dimensions are: $w_1 = w_2 = 3.5$ mm, $l_1 = 5.3$ mm, $h_1 = 16.5$ mm, $h_2 = 5$ mm, $l_{21} = 6.3$ mm, $l_{22} = 8.3$ mm, $L = 50$ mm, and $W = 75$ mm. Owing to the involved geometry of this MMIC

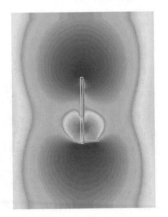

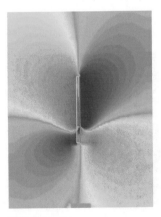

FIGURE 2.21: Snapshots of the E_x, E_y, E_z components at the *y* = 0 plane for the frequency of 2.44 GHz.

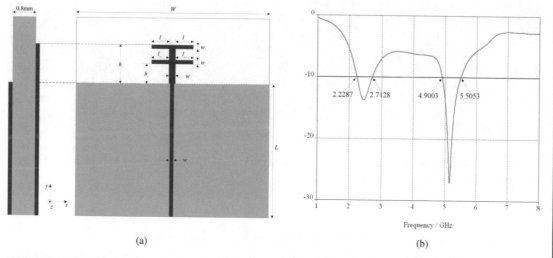

(a)   (b)

FIGURE 2.22: (a) Geometry and (b) reflection coefficient of a dual-band PCB double-monopole T antenna.

antenna, a *hybrid* FDTD/MRTD method at a $160 \times 204 \times 36$ mesh is implemented. In fact, this combination enables the researcher to resolve the highly varying field components in the area of the T arm without excessive resolutions. Hence, Figure 2.22b gives the reflection coefficient of the antenna, pointing out the 484 MHz (from 2.2287 to 2.7128 GHz) bandwidth for the frequency of 2.45 GHz and the 605 MHz one (from 4.9003 to 5.5053 GHz) for that of 5.25 GHz, both at the level of −10 dB. Finally, as before, Figures 2.23–2.25 illustrate the behavior of the three electric field components at different planes through an assortment of time snapshots.

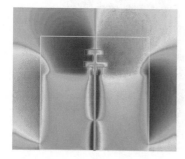
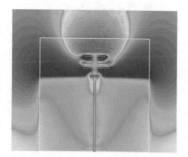

FIGURE 2.23: Snapshots of the E_x, E_y, E_z components at the transverse to the z-axis plane that bisects the substrate for the frequency of 2.45 GHz.

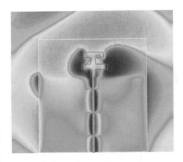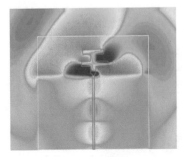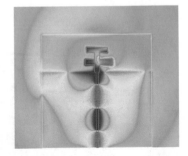

FIGURE 2.24: Snapshots of the E_x, E_y, E_z components at the transverse to the z-axis plane that bisects the substrate for the frequency of 5.25 GHz.

2.8.2 Enhanced Designs With Embedded Circuit Elements

Proceeding to more complex MMIC applications with multiple PCB parts, the circularly polarized antenna of Figure 2.26, fabricated with embedded circuit components, is investigated. The use of the latter along with the incorporation of *two* substrates aims at the improvement of the overall gain and the extension of the total operation bandwidth. The dimensions of the structure are: $D = 30$ mm, $t = 22$ mm, $L_s = 16.4$ mm, $d_s = d_c = 1.2$ mm, whereas the relative permittivities of the substrates are set to $\varepsilon_a = 3.2$ and $c_f = 4.4$, respectively. Due to the geometrical details of the antenna, the traditional FDTD method cannot provide acceptable simulations, unless implementing a very fine,

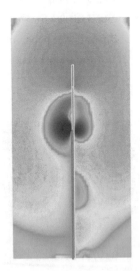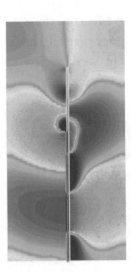

FIGURE 2.25: Snapshots of the E_x, E_y, E_z components at the transverse to x-axis plane that bisects the substrate for the frequency of 5.25 GHz.

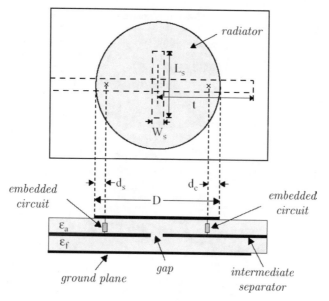

FIGURE 2.26: A circularly polarized MMIC antenna with embedded circuits on multilayered substrate.

yet unaffordable, lattice. To evade this difficulty, the particular device is analyzed through a hybrid FDTD/MRTD and a FDTD/PSTD technique accompanied by an 8 cell PML absorber. In this way, the computational domain is discretized into a grid of $202 \times 186 \times 72$ cells, thus allowing for the satisfactory modeling of the embedded circuits. Figure 2.27 depicts the input impedance (resistance and reactance) of the antenna for two values of the slot dimension W_s. As observed, both hybrid algorithms are in promising agreement with the reference data [33] in the entire range of the frequency spectrum.

Apart from the use of circuit elements and multiple substrates, the performance of an MMIC antenna can be enhanced via thin slits or insets, like the example in Figure 2.28a. The relative permittivity of the single substrate is $\varepsilon_r = 4.4$ and its thickness $h = 1.6$ mm. The rest of the dimensions are set to: $L = 28.5$ mm, $W = 21.8$ mm, $l_s = 2.3$ mm, $d_p = 1.9$ mm, and $W_R = 0.6$ mm. Using the numerical schemes of the previous example, Figure 2.28b shows the 3-dB axial ratio bandwidth as a function of the slit length and the axial ratio frequency f_{AR}. Obviously, the plain FDTD method is proven insufficient for the modeling of such a structure, because its results are rather misleading. On the other hand, both hybrid approaches are, again, found to be very competent and adequately close to the reference data retrieved from [32, 33].

Next, the multiple-microstrip MMIC antenna in Figure 2.29a is examined. Here, the presence of the different microstrips balances the role of the aforesaid thin slits and the embedded

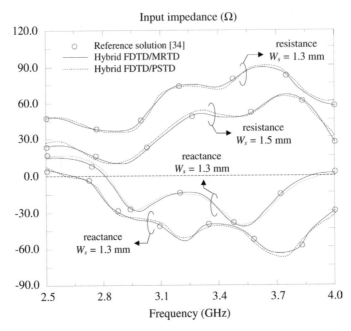

FIGURE 2.27: Input impedance of the circularly polarized MMIC antenna for different slot widths.

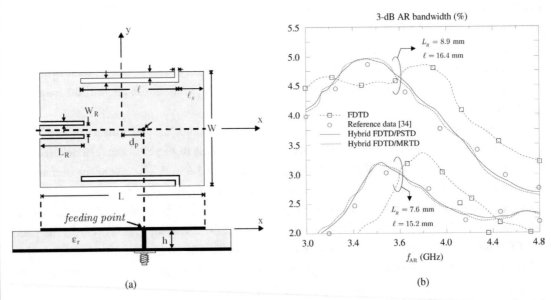

(a) (b)

FIGURE 2.28: (a) Geometry and (b) 3-dB axial ratio bandwidth versus slit length and optimal axial ratio frequency of an MMIC antenna with two slit types.

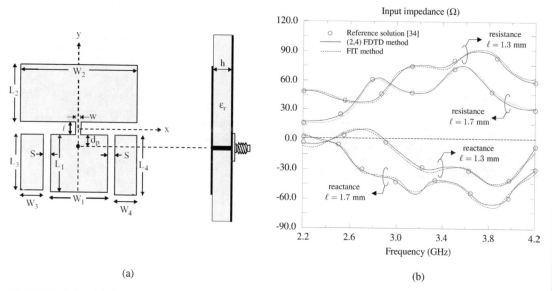

FIGURE 2.29: (a) Geometry and (b) input impedance of an MMIC antenna with multiple microstrip parts.

circuits. Thus, by appropriately adjusting the geometry of each element, one can tune the behavior and the gain of the antenna around a prefixed frequency. For the sake of design convenience, the substrate attributes are kept the same as in Figure 2.28a. A possible set of dimensions could be: $L_1 =$ 10.2 mm, $W_1 = 10.6$ mm, $L_2 = 10.4$ mm, $W_2 = 20$ mm, $L_3 = 9.8$ mm, $W_3 = 3.8$ mm, $L_4 = 10.7$ mm, $W_4 = 3.1$ mm, $S = 1.3$ mm, and $w = 0.5$ mm. The configuration is analyzed by means of the higher-order (2, 4) FDTD method as well as the FIT and the input impedance outcomes for different ℓ values are illustrated in Figure 2.29b. Both techniques are notably rigorous for this type of EMC problem; however, the former requires a more meticulous manipulation of the different media interfaces, where possible instabilities are likely to arise for some distinct cases.

2.9   PCBS IN RADIO FREQUENCY MICROELECTROMECHANICAL SYSTEMS

Microwave MEMS are an emerging field that uses integrated circuit technology or precision mechanical machining on (primarily) silicon wafers in a batch mode to construct microsensors and microactuators for state-of-the-art RF telecommunication arrangements [42–45]. The main advantages of replacing conventional switches and varactors with their MEMS counterparts are cost, size, and weight reduction in addition to an excellent engineering performance—i.e., low losses and high-quality factor—and the reconfigurability they offer. However, entailing the integration of different disciplines such as electrical, mechanical, chemical, or even biological (in their bio-MEMS

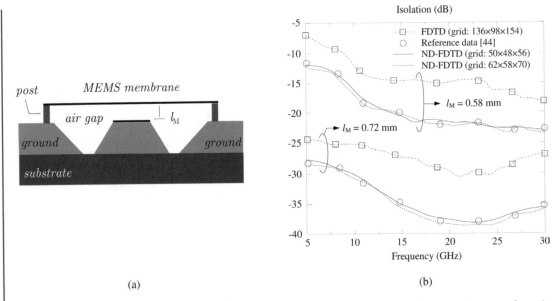

(a)   (b)

FIGURE 2.30: (a) Geometry and (b) isolation of an RF MEMS actuator with a membrane and an air gap.

variant), their compliance with the contemporary international standards should be very carefully examined. In fact, from the EMC designer standpoint, one undesirable effect of RF power on MEMS is intermodulation distortion, which restricts the maximum power a transmitter can deliver due to in-channel and cross-channel interference. Although the levels of this defect are lower than their semiconductor equivalents, this issue should not be ignored.

To experience the importance of the above remarks, let us, first, concentrate on the RF MEMS switch with an undercut profile mounted on an FR4 substrate, as shown in Figure 2.30a. The numerical study is conducted in terms of the nonstandard (ND) FDTD method, which—owing to its advanced discretization features and the highly accurate spatial/temporal operators—requires only $50 \times 48 \times 56$ cells to treat the micron scale, unlike the much larger $136 \times 98 \times 154$ mesh of the simple FDTD approach. Figure 2.30b provides the isolation rate for two l_M values. Notice that the switch copes very well with the intermodulation distortions.

Finally, the 10×12 PCB monolithic antenna of Figure 2.31 is explored. Two types of RF MEMS, namely, power routers and phase shifters, are incorporated, thus augmenting the complication of the structure's computational model. Figure 2.32 illustrates the input impedance and the radiation pattern at 28 GHz of the device for two nonstandard FDTD discretizations. Their superiority over the ordinary FDTD solution is evident, despite the simulation difficulties induced by the notably wideband frequency spectrum.

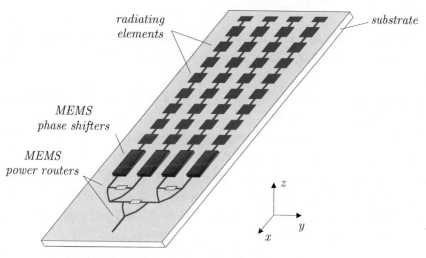

radiating elements

substrate

MEMS phase shifters

MEMS power routers

FIGURE 2.31: A reconfigurable RF MEMS-controlled antenna.

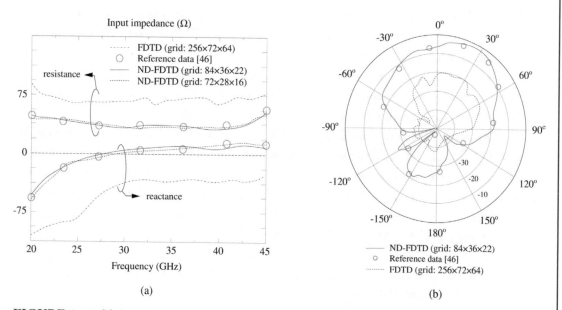

(a)

(b)

FIGURE 2.32: (a) Input impedance and (b) radiation pattern at 28 GHz of the reconfigurable RF MEMS-controlled antenna.

REFERENCES

1. K. S. Yee, "Numerical solution of initial boundary value problems involving Maxwell's equations in isotropic media," *IEEE Trans. Antennas Propagat.*, vol. AP-14, no. 3, pp. 302–307, May 1966.

2. K. S. Kunz and R. J. Luebbers, *The Finite Difference Time Domain Method for Electromagnetics*. Boca Raton, FL: CRC Press, 1993.

3. A. Taflove and S. C. Hagness, *Computational Electrodynamics: The Finite-Difference Time-Domain Method*, 3rd ed. Norwood, MA: Artech House, 2005.

4. P. B. Johns, "A symmetrical condensed node for the TLM-method," *IEEE Trans. Microwave Theory Tech.*, vol. MTT-35, no. 4, pp. 370–377, Apr. 1987. doi:10.1109/TMTT.1987.1133658

5. W. J. Hoefer, "The transmission-line matrix method—Theory and applications," *IEEE Trans. Microwave Theory Tech.*, vol. MTT-33, no. 10, pp. 882–893, Oct. 1995.

6. C. Christopoulos, *The Transmission-Line Modeling (TLM) Method in Electromagnetics*. San Rafael, CA: Morgan & Claypool Publishers, 2006. doi:10.2200/S00027ED1V01Y200605CEM007

7. Q. H. Liu, "The PSTD algorithm: A time-domain method requiring only two cells per wavelength," *Microwave Opt. Technol. Lett.*, vol. 15, no. 3, pp. 158–165, 1997. doi:10.1002/(SICI)1098-2760(19970620)15:3<158::AID-MOP11>3.0.CO;2-3

8. Q. H. Liu and G. Zhao, "Review of the PSTD methods for transient electromagnetics," *Int. J. Numer. Model.: Electron. Netw., Device, Fields*, vol. 22, pp. 299–321, 2004.

9. J. Fan, J. L. Drewniak, H. Shi, and J. Knighten, "DC power-bus modeling and design with a mixed-potential integral-formulation and circuit extraction," *IEEE Trans. Electromagn. Compat.*, vol. 43, no. 4, pp. 426–436, Nov. 2001.

10. Y. Kayano, M. Tanaka, J. L. Drewniak, and H. Inoue, "Common-mode current due to a trace near a PCB edge and its suppression by a guard band," *IEEE Trans. Electromagn. Compat.*, vol. 46, no. 1, pp. 46–53, Feb. 2004. doi:10.1109/TEMC.2004.823609

11. N. V. Kantartzis, T. T. Zygiridis, T. D. Tsiboukis, "An unconditionally stable higher-order ADI-FDTD technique for the dispersionless analysis of generalized 3-D EMC structures," *IEEE Trans. Magn.*, vol. 40, no. 2, pp. 1436–1439, Mar. 2004. doi:10.1109/TMAG.2004.825289

12. M. I. Montrose, E.-P. Li, H.-F. Jin, and W.-L. Yuan, "Analysis on the effectiveness of the 20-H rule for printed-circuit-board layout to reduce edge-radiated coupling," *IEEE Trans. Electromagn. Compat.*, vol. 47, no. 2, pp. 227–233, May 2005. doi:10.1109/TEMC.2005.847383

13. A. P. J. van Deursen and S. Kapora, "Reduction of inductive common-mode coupling of printed circuit boards by nearby U-shaped metal cabinet panel," *IEEE Trans. Electromagn. Compat.*, vol. 47, no. 3, pp. 490–497, Aug. 2005.

14. H.-W. Shim and T. H. Hubing, "Model for estimating radiated emissions from a printed circuit board with attached cables due to voltage-driven sources," *IEEE Trans. Electromagn. Compat.*, vol. 47, no. 4, pp. 899–907, Nov. 2005. doi:10.1109/TEMC.2005.859060

15. J. Kim, M. D. Rotaru, S. Baek, J. Park, M. K. Iyer, and J. Kim, "Analysis of noise coupling from a power distribution network to signal traces in high-speed multilayer printed circuit boards," *IEEE Trans. Electromagn. Compat.*, vol. 48, no. 2, pp. 319–330, May 2006. doi:10.1109/TEMC.2006.873865

16. M. D'Amore and M. S. Sarto, "Theoretical and experimental characterization of the EMP-interaction with composite-metallic enclosures," IEEE Trans. Electromagn. Compat., vol. 42, no. 1, pp. 152–163, Jan. 2000.

17. F. Xiao, W. Liu, and Y. Kami, "Analysis of crosstalk between finite-length microstrip lines: FDTD approach and circuit-concept modelling," *IEEE Trans. Electromagn. Compat.*, vol. 43, no. 4, pp. 573–578, Nov. 2001.

18. V. I. Okhmatovski and A. C. Cangellaris, "A new technique for the derivation of closed-form EM Green's function for unbounded planar layered media," *IEEE Trans. Antennas Propag.*, vol. 50, pp. 1005–1016, Jun. 2002. doi:10.1109/TAP.2002.800731

19. A. G. Polimeridis, T. V. Yioultsis, and T. D. Tsiboukis, "Fast numerical computation of Green's functions for unbounded planar stratified media with a finite-difference technique and Gaussian spectral rules," *IEEE Trans. Microwave Theory Tech.*, vol. 55, no. 1, pp. 100–107, Jan. 2007.

20. C. L. Holloway and E. F. Kuester, "Closed-form expressions for the current densities on the ground planes of asymmetric stripline structures," *IEEE Trans. Electromagn. Compat.*, vol. 49, no. 1, pp. 49–57, Feb. 2007.

21. A. E. Engin, K. Bharath, and M. Swaminathan, "Multilayered finite-difference method (MFDM) for modeling of package and printed circuit board planes," *IEEE Trans. Electromagn. Compat.*, vol. 49, no. 2, pp. 441–447, May 2007.

22. D. E. Anagnostou, M. Morton, J. Papapolymerou, and C. G. Christodoulou, "A 0–55-GHz coplanar waveguide to coplanar strip transition," *IEEE Trans. Microwave Theory Tech.*, vol. 56, no. 1, pp. 1–6, Jan. 2008.

23. S. Grivet-Talocia and F. Canavero, "Wavelet-based high-order, adaptive modeling of lossy interconnects," *IEEE Trans. Electromagn. Compat.*, vol. 43, no. 4, pp. 471–484, Nov. 2001. doi:10.1109/15.974626

24. N. V. Kantartzis, T. I. Kosmanis, T. V. Yioultsis, and T. D. Tsiboukis, "A nonorthogonal higher-order wavelet-oriented FDTD technique for 3-D waveguide structures on generalized curvilinear grids," *IEEE Trans. Magn.*, vol. 37, no. 5, pp. 3264–3268, Sep. 2001. doi:10.1109/20.952591

25. A. Scarlatti and C. L. Holloway, "An equivalent transmission-line model containing dispersion for high-speed digital lines—with an FDTD implementation," *IEEE Trans. Electromagn. Compat.*, vol. 43, no. 4, pp. 504–514, Nov. 2001. doi:10.1109/15.974629

26. M. Leone and V. Navrátil, "On the electromagnetic radiation of printed-circuit-board interconnections," *IEEE Trans. Electromagn. Compat.*, vol. 47, no. 2, pp. 219–226, May 2005. doi:10.1109/TEMC.2005.847400

27. T. Kamgaing and O. M. Ramahi, "Design and modeling of high-impedance electromagnetic surfaces for switching noise suppression in power planes," *IEEE Trans. Electromagn. Compat.*, vol. 47, no. 3, pp. 479–489, Aug. 2005. doi:10.1109/TEMC.2005.850692

28. X. Liu, C. Christopoulos, and D. W. P. Thomas, "Prediction of radiation losses and emission from a bent wire by a network model," *IEEE Trans. Electromagn. Compat.*, vol. 48, no. 3, pp. 476–484, Aug. 2006. doi:10.1109/TEMC.2006.879331

29. D. L. Sounas, N. V. Kantartzis, and T. D. Tsiboukis, "Optimized ADI-FDTD analysis of circularly polarized microstrip and dielectric resonator antennas," *IEEE Microwave Wireless Compon. Lett.*, vol. 16, no. 2, pp. 63–65, Feb. 2006.

30. T. V. Yioultsis, T. I. Kosmanis, I. T. Rekanos, and T. D. Tsiboukis, "EMC analysis of high-speed on-chip interconnects via a mixed quasi-static finite difference-FEM technique," *IEEE Trans. Magn.*, vol. 43, no. 4, pp. 1365–1368, Apr. 2007.

31. Y. Bayram and J. L. Volakis, "A hybrid electromagnetic-circuit method for electromagnetic interference onto mass wires," *IEEE Trans. Electromagn. Compat.*, vol. 49, no. 4, pp. 893–900, Nov. 2007.

32. K. M. Luk and K. W. Leung, Eds., *Dielectric Resonator Antennas*. London, UK: Research Press Studies, 2003.

33. K. L. Wong, *Compact and Broadband Microstrip Antennas*. Piscataway, NJ: Wiley Interscience, 2002. doi:10.1002/0471221112.ch3

34. Z. Li, Y. E. Erdemli, J. L. Volakis and P. Y. Papalambros, "Design optimization of conformal antennas by integrating stochastic algorithms with the hybrid finite-element method," *IEEE Trans. Antennas Propag.*, vol. 50, no. 5, May 2002.

35. J. Gómez-Tagle, P. F. Wahid, M. T. Chryssomallis, and C. G. Christodoulou, "FDTD analysis of finite-sized phased array microstrip antennas, *IEEE Trans. Antennas Propag.*, vol. 51, no. 8, pp. 2057–2062, Aug. 2003. doi:10.1109/TAP.2003.813640

36. N. V. Kantartzis and T. D. Tsiboukis, "A higher order nonstandard FDTD-PML method for the advanced modeling of complex EMC problems in generalized 3-D curvilinear coordinates," *IEEE Trans. Electromagn. Compat.*, vol. 46, pp. 2–11, Feb. 2004. doi:10.1109/TEMC.2004.823606

37. M. N. Vouvakis, C. A. Balanis, C. R. Birtcher, and. A. C. Polycarpou, "Multilayer effects on cavity-backed slot antennas," *IEEE Trans. Antennas Propag.*, vol. 52, no. 3, pp. 880–887, Mar. 2004. doi:10.1109/TAP.2004.824672

38. T. T. Zygiridis, N. V. Kantartzis, and T. D. Tsiboukis, "Higher-order tangential vector finite elements for 3-D antenna array structures," *Electromagn.*, vol. 24, nos. 1–2, pp. 95–111, 2004. doi:10.1080/02726340490261572

39. I. Eshrah, A. A. Kishk, A. B. Yakovlev, and A. W. Glisson, "Theory and implementation of dielectric resonator antenna excited by a waveguide slot," *IEEE Trans. Antennas Propag.*, vol. 53, no. 1, pp. 483–494, Jan. 2005.

40. H.-H. Hsieh, Y.-T. Liao, and L.-H. Lu, "A compact quadrature hybrid MMIC using CMOS active inductors," *IEEE Trans. Microwave Theory Tech.*, vol. 55, no. 6, pp. 1098–1104, Jun. 2007.

41. A. Mehdipour and M. Kamarei, "Fast analysis of external field coupling to orthogonal interconnections in high-speed multiayer MMICs," *IEEE Trans. Electromagn. Compat.*, vol. 49, no. 4, pp. 927–930, Nov. 2007.

42. E. Perret, H. Aubert, and H. Legay, "Scale-changing technique for the electromagnetic modeling of MEMS-controlled planar phase shifters," *IEEE Trans. Microwave Theory Tech.*, vol. 54, no. 9, pp. 3594–3601, Sep. 2006.

43. M. J. Chen, A.-V. H. Pham, N. A. Evers, C. Kapusta, J. Iannotti, W. Kornrumpf, J. J. Maciel, and Karabudal, "Design and development of a package using LCP for RF/microwave MEMS switches," *IEEE Trans. Microwave Theory Tech.*, vol. 54, no. 11, pp. 4009–015, Nov. 2006.

44. W. D. Yuan and R. R. Mansur, "Tunable dielectric resonator bandpass filter with embedded MEMS tuning elements," *IEEE Trans. Microwave Theory Tech.*, vol. 55, no. 1, pp. 154–60, Jan. 2007.

45. B. Lacroix, A. Pothier, A. Crunteanu, C. Cibret, F. Dumas-Bouchiat, C. Champeaux, A. Catherinot, and P. Blondy, "Sub-microsecond RF MEMS switched capacitors," *IEEE Trans. Microwave Theory Tech.*, vol. 55, no. 6, pp. 1314–321, Jun. 2007.

•   •   •   •   •

CHAPTER 3

Electromagnetic Interference, Immunity, Shielding, and Signal Integrity

3.1 INTRODUCTION

Whenever an attempt is made to characterize an EMC event, the physical position of the equipment under study is very likely to play a significant role in the entire procedure, because a noncorrectly resolved location can lead to erroneous estimations regarding the system's endurance to internal or external sources of noise. Actually, the latter disruptions are comprehensively described in the field of EMC engineering through the term electromagnetic interference (EMI). In essence, EMI refers to the mechanism by which disruptive electromagnetic energy propagates from one electronic apparatus to another via radiated or conducted paths. The internal category of EMI is the product of signal degradation along a transmission trail, including parasitic crosstalk among circuits and field coupling between subdevices. Typical remedies focus on confining the periodic signal to as small an area as possible, hence obstructing parasitic paths to the outside world. On the contrary, protection from external disturbances is attained through shielding realizations, appropriately adapted to cover the component's weaknesses. A relative measure of a system's ability to withstand EMI exposure, while preserving a prefixed performance level, is immunity. However, quite often researchers are interested in the tendency of a device to be disrupted by EMI exposure to an incident field, i.e., the lack of immunity or alternatively, its susceptibility. Although several diagnostic techniques have been presented so far, they may be hard to implement, especially when the compatibility issue is amid dissimilar arrangements with unrelated function. For this situation, time-domain computational modeling is probably one of the most reliable simulation tools for immunity and signal integrity testing, as will be deduced from the analysis of this chapter.

3.2 TIME-DOMAIN METHODS FOR EMI CHARACTERIZATION

Apart from the methodologies described in the previous chapter and used herein as well, the next paragraphs give a brief record of the fundamental characteristics of four alternative time-domain algorithms, namely, the FIT, the FETD method, the FVTD algorithm, and the MRTD technique.

3.2.1   The Finite Integration Technique

The FIT establishes its basic notion via a discrete reformulation of Maxwell's equations in their integral version and places six vector components of electric/magnetic voltages and fluxes on a dual lattice [1, 2]. More specifically, the former quantities are assigned to contours, whereas the latter to surfaces, hence allowing for the construction of a dynamic matrix system. The initial stage of the FIT is the determination of the problem, which relates an open region to a bounded domain $\Omega \in \mathbb{R}^3$. To this goal, the method creates the algebraic analogues of Maxwell's equations that guarantee the preservation of *all* physical field properties in the discrete space. Then, Ω is divided into a finite number of N_v cells V_m ($m = 1, 2, ..., N_v$), whose edges and polygonal faces are oriented toward a prefixed direction. This type of decomposition forms G complex, which may be any (non-)rthogonal coordinate mesh. In this work, however, Ω is assumed to comprise bricks and the discretization is attained through the tensor product Cartesian cell complex

$$G \equiv \left\{ V_m \in \mathbb{R}^3 \Rightarrow V_m \equiv [x_i, x_{i+1}] \times \left[y_j, y_{j+1}\right] \times [z_k, z_{k+1}] \right.$$

$$\left. \text{for } i = 1, 2, \ldots, N_x, j = 1, 2, \ldots, N_y, k = 1, 2, \ldots, N_z \right\}, \tag{3.1}$$

where (x_i, y_j, z_k) points are defined by the i, j, k indices along the three axes of the grid. Also in G, the nonempty faces A_m ($m = 1, 2, ..., N_A$) are the intersections of two cells, edges L_m ($m = 1, 2, ..., N_L$) signify the intersection of faces, and nodes P_m ($m = 1, 2, ..., N_P$) are created by the intersection of edges.

Next, Maxwell's laws and the related constitutive equations are discretized by allocating electric voltages, **e**, on the edges and magnetic fluxes, **b**, on the faces of a primary lattice and magnetic voltages, **h**, on the edges and electric fluxes, **d**, on the faces of a secondary (dual) lattice. So, expressing the electric voltage e_m along edge L_m and the corresponding magnetic flux b_m through face A_m, via

$$e_m = \oint_{L_m} \mathbf{E} \cdot d\mathbf{l} \quad \text{and} \quad b_m = \iint_{A_m} \mathbf{B} \cdot d\mathbf{s}, \tag{3.2}$$

Faraday's law in G becomes

$$\oint_{L_m} \mathbf{E} \cdot d\mathbf{l} = -\iint_{A_m} \frac{\partial \mathbf{B}}{\partial t} \cdot d\mathbf{s} \Rightarrow \sum_p c_{mp} e_p = -\frac{\partial b_m}{\partial t}, \tag{3.3}$$

where p is the number of edges required for the circulation. Because the elements are bricks, faces A_m are rectangular and p involves four electric voltages. Furthermore, coefficients c_{mp} contain all the *topological* evidence on the orientation of cell edges within G and hence they exclusively share the values of -1, 0, or $+1$.

On the other hand, Ampère's law is manipulated in a totally analogous way on the secondary complex $\tilde{G}$, created by considering the *foci* of V_m cells as points for the $\tilde{V}_m$ cells of $\tilde{G}$. So, magnetic voltages h_m are defined on the $\tilde{L}_m$ edges of $\tilde{G}$, whereas electric fluxes d_m and current densities j_m on faces $\tilde{A}_m$, in terms of

$$h_m = \oint_{\tilde{L}_m} \mathbf{H} \cdot \mathrm{d}\mathbf{l}, \; d_m = \iint_{\tilde{A}_m} \mathbf{D} \cdot \mathrm{d}s, \; j_m = \iint_{\tilde{A}_m} \mathbf{J} \cdot \mathrm{d}s \qquad (3.4)$$

Therefore, Ampère's law reads

$$\oint_{\tilde{L}_m} \mathbf{H} \cdot \mathrm{d}\mathbf{l} = \iint_{\tilde{A}_m} \left(\frac{\partial \mathbf{D}}{\partial t} + \mathbf{J} \right) \cdot \mathrm{d}s \Rightarrow \sum_p \tilde{c}_{mp} h_p = \frac{\partial d_m}{\partial t} + j_m, \qquad (3.5)$$

with $\tilde{c}_{mp}$ being the counterparts of c_{mp}. Similarly, the remaining Maxwell's equations are transformed to

$$\sum_p s_{mp} b_p = 0 \quad \text{and} \quad \sum_p \tilde{s}_{mp} d_p = q_m, \quad \text{with} \quad q_m = \iiint_{\tilde{V}_m} \rho \mathrm{d}V, \qquad (3.6)$$

s_{mp}, $\tilde{s}_{mp} \in \{-1, 0, +1\}$ the pertinent coefficients for the primary/secondary cell volumes, and p involving six magnetic fluxes for the closed surface integral within volume V_m, due to its hexahedral structure.

The process, prescribed above, yields the following set of explicit matrix expressions,

$$\mathbf{Ce} = -\frac{\mathrm{d}}{\mathrm{d}t} \mathbf{b} \quad \tilde{\mathbf{C}} \mathbf{h} = \frac{\mathrm{d}}{\mathrm{d}t} \mathbf{d} + \mathbf{j}, \qquad (3.7)$$

$$\mathbf{Sb} = 0 \quad \tilde{\mathbf{S}} \mathbf{d} = \mathbf{q}, \qquad (3.8)$$

that contain the evidence from the complex pair $\{G, \tilde{G}\}$ and are known as *Maxwell grid equations (MGE)*. The elements of "curl" matrices $\mathbf{C}$, $\tilde{\mathbf{C}}$ are c_{mp}, $\tilde{c}_{mp}$ and those of "divergence" matrices $\mathbf{S}$, $\tilde{\mathbf{S}}$ are s_{mp}, $\tilde{s}_{mp}$. Regarding the material relation between electric/magnetic voltages and fluxes, the *one-to-one* correspondence of the faces and their penetrating dual edges in Ω, leads to

$$\mathbf{d} = \mathbf{M}_\varepsilon \mathbf{e} + \mathbf{p}, \quad \mathbf{j} = \mathbf{M}_\sigma \mathbf{e}, \quad \mathbf{h} = \mathbf{M}_\mu \mathbf{b} + \mathbf{m}, \qquad (3.9)$$

where $\mathbf{M}_\varepsilon$ and $\mathbf{M}_\mu$ are the positive definite permittivity and permeability matrices, respectively, and $\mathbf{M}_\sigma$ is the semipositive matrix of conductivities. Actually, on Cartesian meshes these sparse matrices have a diagonal profile. Also, $\mathbf{p}$ and $\mathbf{m}$ vectors arise from the existence of permanent electric and magnetic polarizations.

To derive the time-domain FIT expressions, one simply has to use forward and backward finite-difference approximators for the temporal derivatives. Hence, the MGE system is expressed in the compact form of

$$\Xi\mathbf{F} = \Lambda\frac{\partial\mathbf{F}}{\partial t} + \Pi, \tag{3.10}$$

where

$$\Xi = \begin{bmatrix} \mathbf{0} & \tilde{\mathbf{C}} \\ -\mathbf{C} & \mathbf{0} \end{bmatrix}, \quad \Lambda = \begin{bmatrix} \eta_0\mathbf{M}_\varepsilon & \mathbf{0} \\ \mathbf{0} & \eta_0^{-1}\mathbf{M}_\mu \end{bmatrix},$$

and

$$\mathbf{F} = \begin{bmatrix} \eta_0^{-1/2}\mathbf{e} \\ \eta_0^{1/2}\mathbf{h} \end{bmatrix}, \quad \Pi = \begin{bmatrix} \eta_0^{1/2}\mathbf{j} \\ \mathbf{0} \end{bmatrix}, \quad \eta_0 = \sqrt{\frac{\mu_0}{\varepsilon_0}}$$

In this manner, it is easy to acquire the leapfrog integration scheme of the FDTD method. Sampling $\mathbf{b}$ and $\mathbf{e}$ vectors in an interleaving sense with a temporal shift of half a time-step, MGE may be, finally, written as

$$\mathbf{b}^{n+1} = \mathbf{b}^n - \Delta t\mathbf{C}\mathbf{e}^{n+1/2}$$
$$\mathbf{e}^{n+1/2} = \mathbf{e}^{n-1/2} + \Delta t\mathbf{M}_\varepsilon^{-1}\left(\tilde{\mathbf{C}}\mathbf{M}_\mu^{-1}\mathbf{b}^n - \mathbf{j}^n\right) \tag{3.11}$$

in the complex plane and therefore found to be conditionally stable as long as the time-step remains smaller than a prefixed value, determined by a certain criterion. Basically, the most convenient one is the Courant condition explained in Chapter 2. Inspecting the MGE system, it is revealed that its most significant merit is the preservation of all physical properties of the continuous problem. In particular, the resultant degrees of freedom, aggregated in the MGE vectors, are usually measurable integral state variables, whereas the operation matrices are sparse and encompass key data on the incidence relations of the dual grids, so enabling the selection of the suitable time integration process, even for nongauged singular modeling situations.

3.2.2   The Finite-Element Time-Domain Method

Amid various implementations, the method of point-matched finite elements has been the first one to use nodal finite-element ideas for the extraction of explicit schemes for Maxwell's equations [3–5].

Specifically, the algorithm maintains the spatial staggering of electric and magnetic fields; however, all three components of each vector are located at the same node. Thus, two lattices are used for numerical calculations, both built in such a way that every electric field node is enclosed within a cell of the magnetic field grid and every magnetic field node is limited within an element of the electric field grid. This combination leads to the required subspaces for the finite-element approximation, namely,

$$\mathbf{E}(\mathbf{r},t) = \sum_{i=1}^{M_E} \phi_i(\mathbf{r})\mathbf{E}_i(t) \quad \text{and} \quad \mathbf{H}(\mathbf{r},t) = \sum_{j=1}^{M_H} \theta_j(\mathbf{r})\mathbf{H}_j(t), \tag{3.12}$$

with M_E, M_H the number of electric and magnetic field nodes and $\phi_i(\mathbf{r})$, $\theta_j(\mathbf{r})$ scalar basis functions for the accomplishment of rigorous computations in the two meshes, respectively. Plugging (3.12) in Maxwell's laws leads to the following system of state equations

$$\varepsilon_0 \varepsilon_r(\mathbf{r}_i)\frac{d\mathbf{E}_i(t)}{dt} = \sum_{j=1}^{M_H} \nabla\theta_j(\mathbf{r}_i) \times \mathbf{H}_j(t) - \mathbf{J}_c(\mathbf{r}_i), \quad \text{for } i = 1, 2, \ldots, M_E \tag{3.13}$$

$$\mu_0 \mu_r(\mathbf{r}_j)\frac{d\mathbf{H}_j(t)}{dt} = -\sum_{i=1}^{M_E} \nabla\phi_i(\mathbf{r}_j) \times \mathbf{E}_i(t), \quad \text{for } j = 1, 2, \ldots, M, \tag{3.14}$$

To obtain the update FETD expressions, (3.13) and (3.14) must be written in the convenient matrix form of

$$\frac{d\overline{\mathbf{E}}}{dt} = \mathbf{A}^{-1}\left(\mathbf{R}\overline{\mathbf{H}} - \overline{\mathbf{J}}_c\right) \quad \text{and} \quad \frac{d\overline{\mathbf{H}}}{dt} = -\mathbf{C}^{-1}\mathbf{T}\overline{\mathbf{E}}, \tag{3.15}$$

where

$$\overline{\mathbf{E}} = \left[\overline{\mathbf{E}}_1(t), \overline{\mathbf{E}}_2(t), \ldots, \overline{\mathbf{E}}_{M_E}(t)\right]^T, \overline{\mathbf{J}}_c = \left[\overline{\mathbf{J}}_{c,1}(t), \overline{\mathbf{J}}_{c,2}(t), \ldots, \overline{\mathbf{J}}_{c,M_E}(t)\right]^T, \overline{\mathbf{H}} = \left[\overline{\mathbf{H}}_1(t), \overline{\mathbf{H}}_2(t), \ldots, \overline{\mathbf{H}}_{M_H}(t)\right]^T,$$

for

$$\overline{\mathbf{E}}_i(t) = \left[E_{ix}(t), E_{iy}(t), E_{iz}(t)\right]^T, \overline{\mathbf{J}}_{c,i}(t) = \left[J_{c,ix}(t), J_{c,iy}(t), J_{c,iz}(t)\right]^T, \overline{\mathbf{H}}_j = \left[H_{jx}(t), H_{jy}(t), H_{jz}(t)\right]^T$$

$\mathbf{R}$, $\mathbf{T}$ matrices which contain the appropriate curl operations and $\mathbf{A}$, $\mathbf{C}$ matrices with the properties of all materials in the domain. The last step of the method is the approximation of temporal derivatives in (3.15) through a conditionally stable leapfrog-like scheme, so that the temporal increments of $\overline{\mathbf{E}}$ are interleaved one-half time-step with regard to those of $\overline{\mathbf{H}}$. Despite its efficiency, the nodal algorithm suffers from its collocated nature that imposes several setbacks in the derivation of boundary conditions at dissimilar media interfaces.

A robust solution to this problem comes from Whitney elements, whose basis functions are given by

$$\mathbf{w}_{ij}^{(1)} = \zeta_i \nabla \zeta_j - \zeta_j \nabla \zeta_i, \tag{3.16}$$

along edge $\{ij\}$ (where the circulation of (3.16) is equal to unity and zero along all other edges), for ζ_i the linear nodal basis functions connected to vertex i. Hence, any twice-integrable vector field $\mathbf{F}$ may be expressed in terms of Whitney 1-forms as

$$\mathbf{F} = \sum_{\text{edge}\{ij\}} \mathcal{F}_{\text{circ}}^{ij} \mathbf{w}_{ij}^{(1)}, \tag{3.17}$$

with $\mathcal{F}_{\text{circ}}^{ij}$ the circulation of $\mathbf{F}$ along edge $\{ij\}$. Similarly, $\mathbf{F}$, associated to a certain facet $\{ijk\}$, reads

$$\mathbf{F} = \sum_{\text{facet}\{ijk\}} \mathcal{F}_{\text{flux}}^{ijk} \mathbf{w}_{ijk}^{(2)}, \tag{3.18}$$

through Whitney 2-forms, where $\mathbf{w}_{ijk}^{(2)}$ are the corresponding basis functions

$$\mathbf{w}_{ijk}^{(2)} = 2 \left(\zeta_i \nabla \zeta_j \times \nabla \zeta_k + \zeta_j \nabla \zeta_k \times \nabla \zeta_i + \zeta_k \nabla \zeta_i \times \nabla \zeta_j \right), \tag{3.19}$$

and $\mathcal{F}_{\text{flux}}^{ij}$ the flux of $\mathbf{F}$, equal to unity through facet $\{ijk\}$ and zero through all other facets. Selecting, again, a dual-grid tessellation, it is necessary to proceed to the subsequent interpolations of

$$\mathbf{F}_{\text{A}} = \sum_{\text{edge}\{ij\}} \mathcal{F}_{\text{A, circ}}^{ij} \mathbf{w}_{ij}^{(1)} \quad \text{and} \quad \mathbf{F}_{\text{B}} = \sum_{\text{facet}\{ijk\}} \mathcal{F}_{\text{B, flux}}^{ijk} \mathbf{w}_{ijk}^{(2)}, \tag{3.20}$$

for $\mathbf{F}_{\text{A}} = [\mathbf{E}, \mathbf{H}]$, $\mathbf{F}_{\text{B}} = [\mathbf{D}, \mathbf{B}, \mathbf{J}_c]$ and $\mathcal{F}_{\text{A, circ}}^{ij} = \left[\mathcal{E}_{\text{circ}}^{ij}, \mathcal{H}_{\text{circ}}^{ij} \right]$, $\mathcal{F}_{\text{B, circ}}^{ijk} = \left[\mathcal{D}_{\text{flux}}^{ijk}, \mathcal{B}_{\text{flux}}^{ijk}, \mathcal{D}_{\text{c, flux}}^{ijk} \right]$, as extracted from (3.17) and (3.18). Under these abstractions, Ampère's and Faraday's laws become

$$\sum_{\text{edge}\{ij\}} \mathcal{H}_{\text{circ}}^{ij} \nabla \times \mathbf{w}_{ij}^{(1)} = \sum_{\text{facet}\{ijk\}} \left(\frac{\partial \mathcal{D}_{\text{flux}}^{ijk}}{\partial t} + \mathcal{J}_{\text{c, flux}}^{ijk} \right) \mathbf{w}_{ijk}^{(2)}, \tag{3.21}$$

$$\sum_{\text{edge}\{ij\}} \mathcal{E}_{\text{circ}}^{ij} \nabla \times \mathbf{w}_{ij}^{(1)} = - \sum_{\text{facet}\{ijk\}} \frac{\partial \mathcal{B}_{\text{flux}}^{ijk}}{\partial t} \mathbf{w}_{ijk}^{(2)} \tag{3.22}$$

To improve their algebraic profile, (3.21) and (3.22) are processed via the collocation algorithm. For example, on facet $\{ijk\}$ of area Ω_{ijk}, they lead to

$$\iint_{\Omega_{ijk}} \left[\sum_{\text{edge}\{ij\}} \mathcal{H}_{\text{circ}}^{ij} \nabla \times \mathbf{w}_{ij}^{(1)} \right] \cdot d\mathbf{s} = \iint_{\Omega_{ijk}} \left[\sum_{\text{facet}\{ijk\}} \left(\frac{\partial \mathcal{D}_{\text{flux}}^{ijk}}{\partial t} + \mathcal{J}_{\text{c, flux}}^{ijk} \right) \mathbf{w}_{ijk}^{(2)} \right] \cdot d\mathbf{s}, \tag{3.23}$$

$$\iint\limits_{\Omega_{ijk}} \left[\sum_{\text{edge}\{ij\}} \mathcal{E}^{ij}_{\text{circ}} \nabla \times \mathbf{w}^{(1)}_{ij} \right] \cdot d\mathbf{s} = - \iint\limits_{\Omega_{ijk}} \left[\sum_{\text{facet}\{ijk\}} \frac{\partial \mathcal{B}^{ijk}_{\text{flux}}}{\partial t} \mathbf{w}^{(2)}_{ijk} \right] \cdot d\mathbf{s}, \qquad (3.24)$$

and after some calculus to the more general matrix relations of

$$\frac{\partial \mathcal{D}}{\partial t} = \mathbf{Z}\mathcal{H} - \mathcal{J}_{\text{c}} \quad \text{and} \quad \frac{\partial \mathcal{B}}{\partial t} = -\mathbf{Z}\mathcal{E}, \qquad (3.25)$$

where $\mathbf{Z}$ is the circulation array that is filled either by unity or zero entries. For the system of equations to be complete, the next constitutive relations are derived

$$\sum_{\text{facet}\{ijk\}} \mathcal{D}^{ijk}_{\text{flux}} \mathbf{w}^{(2)}_{ijk} = \varepsilon \sum_{\text{edge}\{ij\}} \mathcal{E}^{ij}_{\text{circ}} \mathbf{w}^{(1)}_{ij} \quad \text{and} \quad \sum_{\text{facet}\{ijk\}} \mathcal{B}^{ijk}_{\text{flux}} \mathbf{w}^{(2)}_{ijk} = \mu \sum_{\text{edge}\{ij\}} \mathcal{H}^{ij}_{\text{circ}} \mathbf{w}^{(1)}_{ij}, \qquad (3.26)$$

and in terms of the collocation approach

$$\mathcal{E} = \mathbf{K}^{\text{E}}\mathcal{D} \quad \text{and} \quad \mathcal{H} = \mathbf{K}^{\text{H}}\mathcal{B}, \qquad (3.27)$$

with

$$K^{\text{E}}_{\{ij\}\,\{ijk\}} = \int\limits_{\text{edge}\{ij\}} \varepsilon^{-1} \mathbf{w}^{(2)}_{ijk} \cdot d\mathbf{l} \quad \text{and} \quad K^{\text{H}}_{\{ij\}\,\{ijk\}} = \int\limits_{\text{edge}\{ij\}} \mu^{-1} \mathbf{w}^{(2)}_{ijk} \cdot d\mathbf{l},$$

for $\{ij\}$ and $\{ijk\}$ indices referring to the unknown quantities of edge $\{ij\}$ and facet $\{ijk\}$.

Lastly, if time derivatives in (3.25) are computed by a central finite-difference scheme in consistency with a leapfrog-based rationale, the 3-D update formulae are acquired. Thus,

$$\mathcal{D}^{n+1} = \mathcal{D}^n + \Delta t \mathbf{Z}\mathcal{H}^{n+1/2} - \Delta t \mathcal{J}_{\text{c}}^{n+1/2}, \quad \mathcal{E}^{n+1} = \mathbf{K}^{\text{E}}\mathcal{D}^{n+1}, \qquad (3.28)$$

$$\mathcal{B}^{n+1/2} = \mathcal{B}^{n-1/2} - \Delta t \mathbf{Z}\mathcal{E}^n, \quad \mathcal{H}^{n+1/2} = \mathbf{K}^{\text{H}}\mathcal{B}^{n+1/2}, \qquad (3.29)$$

where Δt is the temporal increment. Their conditional stability is examined via the amplification matrix technique, which, after eliminating $\mathcal{E}$ and $\mathcal{H}$ vectors, concludes to the criterion of

$$\Delta t \leq \frac{2}{\left| \lambda_{\max}\left(\mathbf{Z}\mathbf{K}^{\text{H}}\mathbf{Z}\mathbf{K}^{\text{E}} \right) \right|},$$

where $\lambda_{\max}(\mathbf{N})$ is the largest eigenvalue of matrix $\mathbf{N}$. This limit warrants the choice of an apt Δt, even for the case of involved EMC arrangements with curved boundaries or laborious discontinuities.

3.2.3 The Finite-Volume Time-Domain Algorithm

To overcome the "staircase" limitations of Cartesian meshes, the FVTD technique implements a leapfrog scheme that conserves the total charge both globally and locally [6, 7]. Therefore, the regular staggered FDTD policy is effectively circumvented with the computational burden confined to reasonable levels. The development of the FVTD method can separate the domain into a set of arbitrarily curved dual cells, although the specific paragraph deals with orthogonal discretizations. The corners of each primary cell, whose origin is at (i, j, k) node, are defined as $N_{i,j,k}^{P}$ and those of the secondary cells as $N_{i,j,k}^{S}$. Moreover, assume that the lengths of the matching edges—e.g., the y-directed—are $L_{i,j,k}^{P,2} = \left| N_{i,j+1,k}^{P,y} - N_{i,j,k}^{P,y} \right|$ and $L_{i,j,k}^{S,2} = \left| N_{i,j+1,k}^{S,y} - N_{i,j,k}^{S,y} \right|$, respectively, with $L_{i,j,k}^{P,1}$, $L_{i,j,k}^{P,3}$ and $L_{i,j,k}^{S,1}$, $L_{i,j,k}^{S,3}$ defined in a similar way. Because in general the four corners of each cell face do not always lie on the same plane, it is imperative to describe the geometrical aspects of all edges and faces in both grids. Thus for the former, direction cosines $\mathbf{C}_{i,j,k}^{P,m}$ and $\mathbf{C}_{i,j,k}^{S,m}$, with $m = 1, 2, 3$, of unit vector form are needed. In contrast, primary and secondary faces are typified by their vector areas

$$\mathbf{A}_{i,j,k}^{P,m} = A_{i,j,k}^{P,m} \hat{\mathbf{n}}_{i,j,k}^{P,m} \quad \text{and} \quad \mathbf{A}_{i,j,k}^{S,m} = A_{i,j,k}^{S,m} \hat{\mathbf{n}}_{i,j,k}^{S,m}, \tag{3.30}$$

referring to the $+1$ direction, where $A_{i,j,k}^{P,m}$, $A_{i,j,k}^{S,m}$ are the corresponding scalar areas and $\hat{\mathbf{n}}_{i,j,k}^{P,m}$, $\hat{\mathbf{n}}_{i,j,k}^{S,m}$ the average inward-pointing unit normals of the relative faces. Owing to curvilinear coordinates, $\mathbf{C}_{i,j,k}^{P,m}$ and $\mathbf{C}_{i,j,k}^{S,m}$ will not be mandatorily parallel or perpendicular to $\hat{\mathbf{n}}_{i,j,k}^{P,m}$ and $\hat{\mathbf{n}}_{i,j,k}^{S,m}$. Such an issue implies that the 3×3 matrices created by the dot products of $\hat{\mathbf{n}}_{i,j,k}^{S,m}$ and $\mathbf{C}_{i,j,k}^{P,m}$ or vice versa will not be unitary. Thus, the diagonal elements of these matrices must be obtained in a preprocessing stage. For instance,

$$\hat{\mathbf{n}}_{i,j,k}^{S,1} \cdot \mathbf{C}_{i,j,k}^{P,1} = n_{i,j-1,k-1}^{S,1x} C_{i,j,k}^{P,1x} + n_{i,j-1,k-1}^{S,1y} C_{i,j,k}^{P,1y} + n_{i,j-1,k-1}^{S,1z} C_{i,j,k}^{P,1z}, \tag{3.31}$$

with equivalent relations holding for $\hat{\mathbf{n}}_{i,j,k}^{S,2} \cdot \mathbf{C}_{i,j,k}^{P,2}$ and $\hat{\mathbf{n}}_{i,j,k}^{S,3} \cdot \mathbf{C}_{i,j,k}^{P,3}$ as well as for the $\hat{\mathbf{n}}_{i,j,k}^{P,m} \cdot \mathbf{C}_{i,j,k}^{S,m}$ elements.

Substitution of the previous quantities to Maxwell's laws, for each primary face, gives

$$\mu_0 \iint_{A_{i,j,k}^{P,m}} \frac{\partial \mathbf{H}}{\partial t} \cdot \hat{\mathbf{n}}_{i,j,k}^{P,m} dA = - \oint_{\partial A_{i,j,k}^{P,m}} \mathbf{E} \cdot d\mathbf{l}, \tag{3.32}$$

which presumes the projection of $\mathbf{E}$ along the cell edges and simultaneously provides the projection of $\mathbf{H}$ toward the face unit vectors. Because (3.32), through the Stokes theorem, stipulates the $\mathbf{H} \cdot \mathbf{C}_{i,j,k}^{S,m}$ term for the update of $\mathbf{E}$, one must calculate the following relation

$$\mathbf{C}_{i,j,k}^{S,m} = \left(\mathbf{C}_{i,j,k}^{S,m} \cdot \hat{\mathbf{n}}_{i,j,k}^{P,m} \right) \hat{\mathbf{n}}_{i,j,k}^{P,m} + \left(\mathbf{C}_{i,j,k}^{S,m} \cdot \hat{\mathbf{n}}_{i,j,k}^{P,m+1} \right) \hat{\mathbf{n}}_{i,j,k}^{P,m+1} + \left(\mathbf{C}_{i,j,k}^{S,m} \cdot \hat{\mathbf{n}}_{i,j,k}^{P,m+2} \right) \hat{\mathbf{n}}_{i,j,k}^{P,m+2}, \tag{3.33}$$

with the $m + 1$ and $m + 2$ standing for the cyclic permutation of 1, 2, 3. In this framework, (3.33) yields

$$\mu_0 A_{i,j,k}^{P,m} \frac{\partial \left(\mathbf{H} \cdot \mathbf{C}_{i,j,k}^{S,m} \right)}{\partial t} = \mu_0 A_{i,j,k}^{P,m} \left[\frac{\partial \left(\mathbf{H} \cdot \hat{\mathbf{n}}_{i,j,k}^{P,m} \right)}{\partial t} \left(\hat{\mathbf{n}}_{i,j,k}^{P,m} \cdot \mathbf{C}_{i,j,k}^{S,m} \right) \right.$$

$$\left. + \frac{\partial \left(\mathbf{H} \cdot \hat{\mathbf{n}}_{i,j,k}^{P,m+1} \right)}{\partial t} \left(\hat{\mathbf{n}}_{i,j,k}^{P,m+1} \cdot \mathbf{C}_{i,j,k}^{S,m} \right) + \frac{\partial \left(\mathbf{H} \cdot \hat{\mathbf{n}}_{i,j,k}^{P,m+2} \right)}{\partial t} \left(\hat{\mathbf{n}}_{i,j,k}^{P,m+2} \cdot \mathbf{C}_{i,j,k}^{S,m} \right) \right] \qquad (3.34)$$

Both first- and second-order patterns can be used for derivative approximation in (3.34). According to the former, the last two terms in the right-hand side of the equation are ignored. Despite the lower accuracy in some EMC problems than their second-order constituents, their simplicity is frequently preferable.

The first-order approximation of (3.34) is

$$\mu_0 A_{i,j,k}^{P,m} \frac{\partial \left(\mathbf{H} \cdot \mathbf{C}_{i,j,k}^{S,m} \right)}{\partial t} = - \left(\hat{\mathbf{n}}_{i,j,k}^{P,m} \cdot \mathbf{C}_{i,j,k}^{S,m} \right) \oint_{\partial A_{i,j,k}^{P,m}} \left(\mathbf{E} \cdot \mathbf{C}_{i,j,k}^{P,m} \right) dl, \qquad (3.35)$$

with the respective and completely dyadic expression for the magnetic field curl given by

$$\varepsilon_0 A_{i,j,k}^{S,m} \frac{\partial \left(\mathbf{E} \cdot \mathbf{C}_{i,j,k}^{P,m} \right)}{\partial t} = \left(\hat{\mathbf{n}}_{i,j,k}^{S,m} \cdot \mathbf{C}_{i,j,k}^{P,m} \right) \oint_{\partial A_{i,j,k}^{S,m}} \left(\mathbf{H} \cdot \mathbf{C}_{i,j,k}^{S,m} \right) dl \qquad (3.36)$$

The discretization of (3.35), for $m = 1$, leads to

$$\mu_0 A_{i,j,k}^{P,1} \frac{\mathbf{H}_{i,j,k}^{n+1/2} \cdot \mathbf{C}_{i,j,k}^{S,1} - \mathbf{H}_{i,j,k}^{n-1/2} \cdot \mathbf{C}_{i,j,k}^{S,1}}{\Delta t} = - \left(\hat{\mathbf{n}}_{i,j,k}^{P,1} \cdot \mathbf{C}_{i,j,k}^{S,1} \right) \left[\left(\mathbf{E}_{i,j+1,k}^{n} \cdot \mathbf{C}_{i,j+1,k}^{P,3} \right) L_{i,j+1,k}^{P,3} \right.$$

$$- \left(\mathbf{E}_{i,j,k}^{n} \cdot \mathbf{C}_{i,j,k}^{P,3} \right) L_{i,j,k}^{P,3} - \left(\mathbf{E}_{i,j,k+1}^{n} \cdot \mathbf{C}_{i,j,k+1}^{P,2} \right) L_{i,j,k+1}^{P,2}$$

$$\left. + \left(\mathbf{E}_{i,j,k}^{n} \cdot \mathbf{C}_{i,j,k}^{P,2} \right) L_{i,j,k}^{P,2} \right], \qquad (3.37)$$

together with two supplementary equations for $\mathbf{H} \cdot \mathbf{C}_{i,j,k}^{S,2}$ and $\mathbf{H} \cdot \mathbf{C}_{i,j,k}^{S,3}$. Similar 3-D FVTD forms are readily derived for the temporal update of $\mathbf{E} \cdot \mathbf{C}_{i,j,k}^{P,1}$, $\mathbf{E} \cdot \mathbf{C}_{i,j,k}^{P,2}$, and $\mathbf{E} \cdot \mathbf{C}_{i,j,k}^{P,3}$ from (3.36).

3.2.4   The Multiresolution Time-Domain Technique

The development of the MRTD method is founded on the representation of electromagnetic fields via scaling and wavelet functions that offer variable grid gradings [8–10]. In particular, the expansion in terms of the former only is deemed ideal for the analysis of smooth quantities, whereas the combination of both counterparts is important for areas with point singularities. The schemes are devised by spatially truncated cubic Battle-Lemarie or Haar scaling and wavelet orthonormal functions. Thus, the resulting technique is consistent and attains accurate outcomes for discrete models near the Nyquist sampling condition. Principally, the general formulation of the MRTD schemes stems from

$$g(s) = \sum_l a_l \phi(s-l) + \sum_m \sum_l b_{ml} \psi(2^m s - l), \tag{3.38}$$

where $g(s)$ is a real twice-integrable function and a_l, b_{ml} are weighting parameters. In (3.38), the first sum is the projection of $g(s)$ onto subspace G_0. The basis of G_0 is acquired through orthogonal translations of scaling functions $\phi(s)$. Then, the resolution of G_0 is refined by the second sum of (3.38), which gives the projections of $g(s)$ onto subspaces L_m, each defined by a wavelet basis $\{\psi(2^m - l)\}$. Function $\psi_{m,l}(s)$ is called the mother wavelet, as all other wavelet members are produced by recursive dilations of $\psi(s)$. Selecting the Haar basis, which is the most pertinent for the interpretation of multiresolution notions,

$$\phi(s) = \begin{cases} 1, & 0 \le s < 1 \\ 0, & \text{otherwise} \end{cases} \tag{3.39}$$

in the form of $\phi_l(s) = (s-l)$, with the mother wavelet function, for normalized s variables, given by

$$\psi(s) = \begin{cases} 1, & 0 \le s < 1/2 \\ -1, & 1/2 \le s < 1 \\ 0, & \text{otherwise} \end{cases} \tag{3.40}$$

So, the desired wavelet basis is created by means of translations and dilations, as

$$\psi_{m,l}(s) = \sqrt{2^m}\, \psi(2^m s - l), \tag{3.41}$$

a universal process for any dyadic set of functions. Now, the extraction of the MRTD schemes requires an additional term in field expansions, for space representation, weighted by a prefixed wavelet set. Taking into account that higher dimensionalities are also easily supported, the analysis at hand focuses on the y-directed case. Therefore, using (3.38), electric and magnetic field components read

$$E_x(x,y,z,t) = \sum_{i,j,k,n=-\infty}^{+\infty} \left[E_x|_{i+1/2,j,k}^n q_j(y) + \bar{E}_x|_{i+1/2,j+1/2,k}^n \psi_{j+1/2}(y) \right] Y_{i+1/2,k}^n, \tag{3.42}$$

$$E_y(x,y,z,t) = \sum_{i,j,k,n=-\infty}^{+\infty} \left[E_y|_{i,j+1/2,k}^n q_{j+1/2}(y) + \bar{E}_y|_{i,j,k}^n \psi_j(y) \right] Y_{i,k}^n, \tag{3.43}$$

$$E_z(x,y,z,t) = \sum_{i,j,k,n=-\infty}^{+\infty} \left[E_z\big|_{i,j,k+1/2}^{n} q_j(y) + \overline{E}_z\big|_{i,j+1/2,k+1/2}^{n} \psi_{j+1/2}(y) \right] Y_{i,k+1/2}^{n}, \quad (3.44)$$

$$H_x(x,y,z,t) = \sum_{i,j,k,n=-\infty}^{+\infty} \left[H_x\big|_{i,j+1/2,k+1/2}^{n+1/2} q_{j+1/2}(y) + \overline{H}_x\big|_{i,j,k+1/2}^{n+1/2} \psi_j(y) \right] Y_{i,k+1/2}^{n+1/2}, \quad (3.45)$$

$$H_y(x,y,z,t) = \sum_{i,j,k,n=-\infty}^{+\infty} \left[H_y\big|_{i+1/2,j,k+1/2}^{n+1/2} q_j(y) + \overline{H}_y\big|_{i+1/2,j+1/2,k+1/2}^{n+1/2} \psi_{j+1/2}(y) \right] Y_{i+1/2,k+1/2}^{n+1/2}, \quad (3.46)$$

$$H_z(x,y,z,t) = \sum_{i,j,k,n=-\infty}^{+\infty} \left[H_z\big|_{i+1/2,j+1/2,k}^{n+1/2} q_{j+1/2}(y) + \overline{H}_z\big|_{i+1/2,j,k}^{n+1/2} \psi_j(y) \right] Y_{i+1/2,k}^{n+1/2}, \quad (3.47)$$

where the bar over E, H signifies the extra expansion coefficients and $Y_{i,k}^{n} = q_i(x)q_k(z)p_n(t)$, with $q_m(u)$

$$q_m(t) = Q\left(\frac{u}{\Delta u} - m\right), \quad (3.48)$$

for $m = i, j, k$; $u = x, y, z$; Δu is the spatial increment; and $Q(u)$ is the cubic spline Battle-Lemarie scaling function. Moreover, $p_n(t)$ is denoted by

$$p_n(t) = P\left(\frac{t}{\Delta t} - n\right) \quad \text{in which} \quad P(t) = \begin{cases} 1 & \text{for } |t| < 1/2 \\ 1/2 & \text{for } |t| = 1/2 \\ 0 & \text{for } |t| > 1/2 \end{cases} \quad (3.49)$$

On the other hand, $\psi_{m+1/2}(y)$ is evaluated through

$$\psi_{m+1/2}(y) = \psi\left(\frac{y}{\Delta y} - m\right), \quad (3.50)$$

with $\psi(y)$ the Battle-Lemarie wavelet function. Note that although $q_m(u)$ and $p_n(t)$ are evenly symmetric with regard to $x = 0$, function $\psi(y)$ has an even symmetry with reference to $x = 1/2$. To include the temporal resolution level in (3.50), wavelet function

$$\psi_{n,m+1/2}(y) = \sqrt{2^n}\,\psi\left(2^n\frac{y}{\Delta y} - m\right), \quad (3.51)$$

is considered a very viable choice. Thus, the resulting algorithm yields precise solutions with adequate dispersion error traits. In the case of large frequency spectra, the latter errors are virtually trivial without any deviations from the expected behavior. However, in real-world EMC problems, where higher modes coexist, the MRTD realization opts for supplementary wavelets along the remaining directions in the grid.

Before plugging (3.42)–(3.47) into Maxwell's equations, a set of orthogonality relations should be applied. Such a requisite, after the necessary algebra, leads to the numerical computation of

$$\int_{-\infty}^{+\infty} \psi_l(y) \frac{\partial \phi_m(y)}{\partial y} dy \simeq \sum_{\tau=-9}^{9} \vartheta(\tau) \delta_{l+\tau,m}, \tag{3.52}$$

$$\int_{-\infty}^{+\infty} \phi_l(y) \frac{\partial \psi_{m+1/2}(y)}{\partial y} dy \simeq \sum_{\tau=-9}^{9} \vartheta(\tau) \delta_{l+\tau,m+1}, \tag{3.53}$$

$$\int_{-\infty}^{+\infty} \psi_l(y) \frac{\partial \psi_{m+1/2}(y)}{\partial y} dy \simeq \sum_{\tau=-9}^{8} \zeta(\tau) \delta_{l+\tau,m}, \tag{3.54}$$

with $\vartheta(\tau)$ and $\zeta(\tau)$ real coefficients. The truncation in the sums of (3.52)–(3.54) is accredited to the exponentially decaying content of both $q_m(u)$ and $\psi_{m+1/2}(y)$ that render the $\tau > 9$ terms of minor interest. After the above issues, Maxwell's equations can be easily sampled in space and time. As an illustration, the x-component of Ampère's law receives the form of

$$\frac{\varepsilon}{\Delta t} \left(E_x|_{i+1/2,j,k}^{n+1} - E_x|_{i+1/2,j,k}^{n} \right) = \sum_{\tau=-9}^{8} \xi(\tau) \left[\frac{1}{\Delta y} H_z|_{i+1/2,j+\tau+1/2,k}^{n+1/2} - \frac{1}{\Delta z} H_y|_{i+1/2,j,k+\tau+1/2}^{n+1/2} \right]$$

$$+ \frac{1}{\Delta y} \sum_{\tau=-9}^{9} \vartheta(\tau) \bar{H}_z|_{i+1/2,j+\tau,k}^{n+1/2} \tag{3.55}$$

$$\frac{\varepsilon}{\Delta t} \left(\bar{E}_x|_{i+1/2,j+1/2,k}^{n+1} - E_x|_{i+1/2,j+1/2,k}^{n} \right) = \sum_{\tau=-9}^{8} \left[\frac{\zeta(\tau)}{\Delta y} \bar{H}_z|_{i+1/2,j+\tau+1,k}^{n+1/2} \right.$$

$$\left. - \frac{\xi(\tau)}{\Delta z} \bar{H}_y|_{i+1/2,j+1/2,k+\tau+1/2}^{n+1/2} \right]$$

$$+ \frac{1}{\Delta y} \sum_{\tau=-9}^{9} \vartheta(\tau) H_z|_{i+1/2,j+\tau+1/2,k}^{n+1/2} \tag{3.56}$$

3.3 EMI ANALYSIS AND IMMUNITY TESTING OF ANTENNA STRUCTURES

This section investigates the impact of EMI from different sources on the proper operation of antennas and performs the required immunity testing simulations. The particular type of EMC problems presents a number of nontrivial difficulties, such as the arbitrary interactions between the radiated wave fronts and the parasitic modes as well as the complexity of the computational model due to realistic details [11–34]. For this purpose, the examples are simulated via various time-domain techniques to help the researcher experience the pros and the cons of each algorithm and decide on its potential usage according to the application under study.

3.3.1 Enhanced-Performance Radiators With a Complex Geometry

The first application for EMI analysis is the two-arm, conical spiral antenna (CSA), shown in Figure 3.1. The CSA is a well-known broadband radiator that has several desirable features when isolated in free space, such as uniform input impedance, large gain, and circular polarization [20, 21]. Actually, its operation is based on two necessary conditions for structures of this type: the angle principle and the truncation principle. According to the former, the performance of an antenna, defined completely by angles, will be frequency-independent. Such antennas are infinite in size, hence an extra convention is required for practical realizations. The latter principle states that the antenna must have an "active region" of finite size that is responsible for the radiation at a particular frequency. As the frequency changes, this region moves on the radiator, so that its dimensions, expressed in terms of the wavelength, remain constant; consequently, the overall performance remains constant. In this manner, the antenna is then practically frequency-independent over the range of frequencies for which the active region is entirely contained within the finite structure of the antenna.

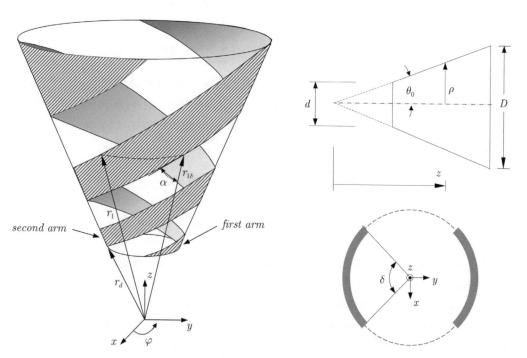

FIGURE 3.1: Geometry of the two-arm CSA as well as side and top views of its basic cone.

Nonetheless, intended to work isolated in free space, CSA is rather vulnerable to EMI exposure caused by several sources, not necessarily in relevance with its close environment.

This is exactly the reason for its examination by means of time-domain computational modeling. Thus, and before any numerical outcomes, it is deemed instructive to provide a brief description of the antenna's major features. In essence, CSA is constructed by winding two metallic strips around the surface of a truncated cone. The whole geometry is defined by the half angle of the cone, θ_0, the wrap angle α, and the angular width of the arms, δ. When θ_0 is small, the CSA radiation is primarily along the axis of the cone, whereas for $\theta_0 = 90°$ the conical spiral becomes a planar spiral, which radiates equally in two directions. Moreover, angle δ denotes the constant angular width of the arms everywhere along the cone, as illustrated in Figure 3.1. Typically, the most usual setup is that of $\delta = 90°$, where *both* metallic arms are equal in size and shape to the open regions on the conical surface. The extent of such an antenna is confined by the minimum and maximum diameters of the cone, d and D, respectively.

Concerning the mathematical description of the CSA, its arm boundaries can be expressed in terms of the radial distance r. Thus, the corresponding quantities r_1 and $r_{1\delta}$ from Figure 3.1, are given by

$$r_1(\varphi) = r_d e^{b|\varphi|}, \quad \text{for} \quad |\varphi| \geq 0, \tag{3.57}$$

$$r_{1\delta}(\varphi) = r_d e^{b(|\varphi|-\delta)}, \quad \text{for} \quad |\varphi| \geq \delta, \tag{3.58}$$

where $b = \sin \theta_0 / \tan \alpha$ and $r_d = d/(2 \sin \theta_0)$ is the radial distance from the base (apex) to the smaller end of the cone (diameter d). The two arms are *symmetric* to the z-axis (diametrically opposite), so the boundaries of the second arm can be obtained by rotating the boundaries of the first arm by angle π, namely,

$$r_2(\varphi) = r_d e^{b(|\varphi|-\pi)}, \quad \text{for} \quad |\varphi| \geq \pi, \tag{3.59}$$

$$r_{2\delta}(\varphi) = r_d e^{b(|\varphi|-\pi-\delta)}, \quad \text{for} \quad |\varphi| \geq \pi + \delta, \tag{3.60}$$

Because of the multiple curvatures encountered in the body of the CSA, its numerical discretization is performed via the conformal FDTD technique, the FVTD algorithm, and the FIT, i.e., methods with enhanced competences in the manipulation of arbitrary geometries. In the FDTD model, in particular, space is divided into cubical cells and the CSA perfect electrically conducting (PEC) arms are approximated either by a staircase or a conformal surface. This surface is formed using the two-stage process of Figure 3.2a, where the solid circle represents the intersection of the conical CSA surface with the z-plane and the dashed one the intersection of the conical surface with the plane one cell below $(z - \Delta z)$. During the first stage, the circular boundary of the antenna is modeled by vertical PEC faces (thick lines), whose projection above the plane is at the solid circle and below the plane at the dashed circle. On the plane of the figure, these vertical faces are connected by a horizontal PEC surface (shaded area). In the second stage, given in Figure 3.2b, the angular sectors of width δ are determined (thick lines), and the horizontal/vertical faces that are within these sectors are extracted to form the arms on this plane. The above process is repeated on consecutive z-planes until the complete antenna is obtained. As for the excitation of the CSA, the incident voltage differentiated Gaussian pulse

$$V_{in}(t) = -\frac{V_0 t}{\tau} e^{0.5\left[1-(t/\tau)^2\right]} \tag{3.61}$$

in the feeding transmission line (Figure 3.3) excites the structure. In (3.61), τ is the characteristic time of the pulse, which has the asset that it does not contain a zero-frequency component and thus avoids prolonged simulations. It is important to mention that the EMI sources for the immunity study in the surroundings of the antenna are realized through the well-known total-/scattered-field formulation, whereas the computational domain is terminated by means of an eight-cell PML absorber, adjusted to the requirements of each numerical technique.

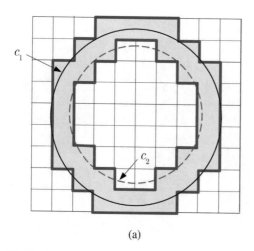

(a)

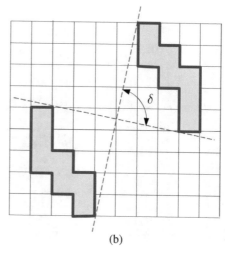

(b)

FIGURE 3.2: Graphical depiction of the FDTD modeling of the CSA arms in a two-stage procedure.

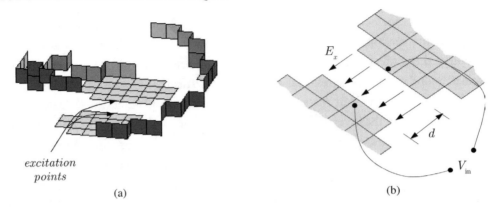

(a)   (b)

FIGURE 3.3: (a) The feeding line of the CSA in the discretized domain and (b) detail of the excitation points.

Proceeding to the results, Figure 3.4 gives the snapshots of E_x component at different time-steps. It is easily observed that the electromagnetic energy is gradually radiated, whereas the imposed EMI does affect the phenomenon by hindering its normal evolution. More specifically, the excitation pulse moves along the CSA length, hence letting the higher frequencies to appear first and the lower ones to emerge next.

Next, the CSA radiation performance and beam directivity capabilities are examined in the presence of a strong EMI status. In this context, Figures 3.5 and 3.6 show its radiation patterns at the $\varphi = 0°$ plane for $0° \leq \theta \leq 180°$ with $\theta_0 = 7.5°$, $\delta = 90°$, $D/d = 8$, and $\alpha = 40°, 55°$, respectively.

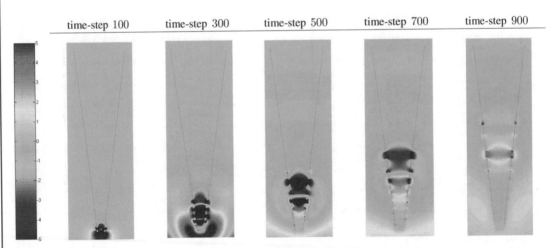

FIGURE 3.4: Snapshots of the E_x component at the $y = 0$ plane for a two-arm CSA with $\theta_0 = 7.5°$, $\alpha = 55°$, $\delta = 90°$, and $D/d = 8$.

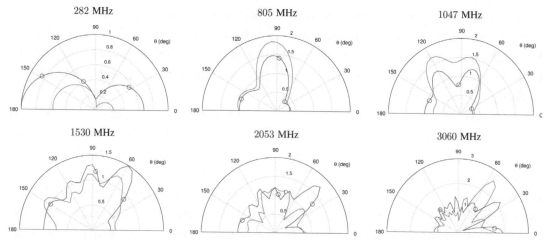

FIGURE 3.5: Radiation patterns of a two-arm CSA with $\theta_0 = 7.5°$, $\alpha = 40°$, $\delta = 90°$, and $D/d = 8$ at diverse frequencies for $\varphi = 0°$ and $0 \leq \theta \leq 180°$. The straight line refers to the E_φ and the circled one to the E_q component.

Inspecting the diagrams, one may come up with the conclusion that for both α values, there exists an optimal frequency range of operation where the electromagnetic energy propagates along the $-z$ direction. In other words, at these frequencies the main lobe is formed with its maximum toward $-z$. For lower frequencies, on the other hand, although this lobe is still present, it is not the

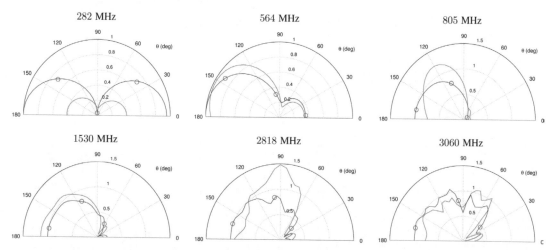

FIGURE 3.6: Radiation patterns of a two-arm CSA with $\theta_0 = 7.5°$, $\alpha = 55°$, $\delta = 90°$, and $D/d = 8$ at diverse frequencies for $\varphi = 0°$ and $0 \leq \theta \leq 180°$. The straight line refers to the E_φ and the circled one to the E_q component.

dominant anymore, because a significant portion of energy is radiated along the +z axis. Moreover, it is instructive to stress that at high frequencies the *half-power beamwidth (HPBW)* of the aforesaid lobe is *seriously* increased, while a set of nontrivial side lobes appears. Similarly, Figure 3.7 reveals the close dependence of the CSA operation characteristics on angles α and θ_0, which determine its geometry.

In particular, the figure provides the radiation patterns at the $\varphi = 0°$ plane for four distinct frequencies and variable α and θ_0. From the results, it becomes apparent as α tends to 90° and the EMI magnitude gets stronger, the radiated energy of the antenna travels along the $\theta_0 = 180°$ direction. Thus, at the frequency of 520 MHz, where a frontal lobe is always present, the HPBW is smaller for $\alpha = 55°$ and 65°. However, at higher frequencies various side lobes that degrade the CSA directivity start to appear. Equivalent deductions may be extracted from the fluctuation of θ_0, with the most notable one being the *improvement* of the antenna performance as the specific angle receives smaller values.

Apart from the influence of EMI exposure on the immunity traits of the CSA, its role in the variation of two other key quantities should be examined. These are *front-to-back ratio*

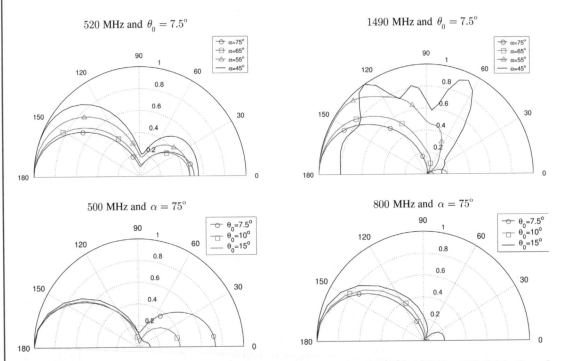

FIGURE 3.7: Radiation patterns for the E_θ component of a two-arm CSA with d = 90°, $D/d = 8$, and variable α and $\theta_{\text{à}}$.

(FTBR) and the already mentioned HPBW [19, 20]. The former indicator (expressed in dB) is defined as the ratio of the emitted power along the main direction of radiation to the emitted power toward the opposite direction

$$\text{FTBR} = 20 \log_{10} \left| E_i^{\text{rad}} \left(\theta = 180°, \varphi = 0° \right) \right| - 20 \log_{10} \left| E_i^{\text{rad}} \left(\theta = 0°, \varphi = 0° \right) \right|, \quad (3.62)$$

where index i can stand for θ or φ components. Figure 3.8a illustrates the FTBR of a CSA with $\theta_0 = 7.5°$, $\delta = 90°$, $d/D = 8$, and variable α.

Note that the FTBR augments as α increases, everywhere in the frequency spectrum, so extending the optimal operation range of the antenna. Such a behavior should be anticipated, because large α values *enhance* the CSA overall performance; an issue also substantiated by the corresponding radiation patterns of Figure 3.7. Conversely, the HPBW indicator is defined as the angle between two symmetric points of the main lobe where the magnitude of electric field intensity becomes equal to 0.707 of its maximum value. Figure 3.8b displays a set of interesting outcomes for different angles α, which indicate a clear *improvement* (i.e., reduction) of HPBW with the increase of α.

Finally, Figure 3.9 provides the axial ratio of the CSA versus frequency when $\theta_0 = 10.5°$ and $\alpha = 75°$. From the results, one can promptly discern the relatively large frequency range (from 800 MHz to 2.7 GHz) where this ratio remains below 1.1, implying that the polarization of the antenna along the $-z$ axis (maximum gain direction) is very close to the circular type.

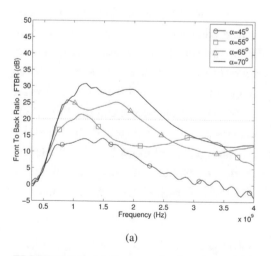

(a)

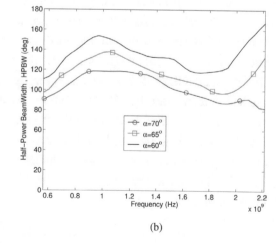

(b)

FIGURE 3.8: (a) Front-to-back ratio and (b) half-power beamwidth as a function of frequency for a two-arm CSA with $\theta_0 = 7.5°$, $\delta = 90°$, $D/d = 8$, and variable α.

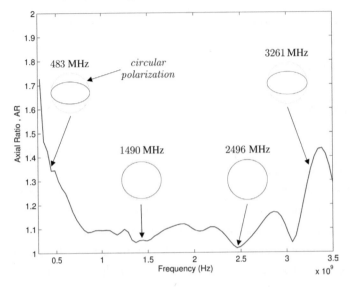

FIGURE 3.9: Axial ratio as a function of frequency of a two-arm CSA with $\theta_0 = 7.5°$, $\alpha = 70°$, $\delta = 90°$, and $D/d = 8$. Four distinct polarizations are also presented, where the dashed line indicates the circular polarization with a diameter equal to the large axis of the corresponding ellipsis.

The next type of radiator that will be investigated from an immunity point of view is the semiconical substrate antenna in Figure 3.10. In its general form, this antenna consists of microstrips appropriately mounted on a dielectric semiconical substrate over an infinite ground plane. Owing to the latter attribute, it can be profitably used in PCB components for several microwave applications. The consecutive parallel microstrips are connected via thin metallic wires and the basic design parameters are: the half-span angle θ_0, the microstrip angular span ψ, and width t_s as well as the intermediate distance t_d between two neighboring microstrips. In addition, the cone bases are described by radii R_1 and R_2, whereas of critical significance for the performance of the structure is the substrate relative dielectric permittivity ε_r. The antenna is numerically simulated by means of a hybrid time-domain method that employs the FVTD technique for the curved parts of the device and the FDTD algorithm for the rest of the domain. For the sake of comparison, a combined second-/higher-order FDTD approach is also used. Although various lattices and grid resolutions have been selected depending on the case under study, an indicative lattice for the prior antenna comprises $28 \times 56 \times 114$ cells. The whole arrangement, truncated by a six-cell PML, is excited through the modulated Gaussian pulse of

$$V_{in}(t) = V_0 e^{-(t/\tau)} \sin(2\pi f_0 t), \qquad (3.63)$$

where τ is the characteristic time of the pulse and f_0 is the central frequency of the spectrum.

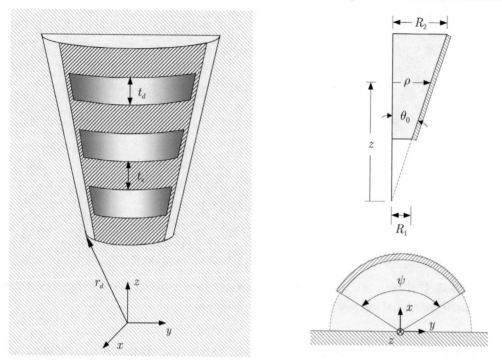

FIGURE 3.10: Geometry of the semiconical substrate antenna as well as side and frontal view of its basic cone.

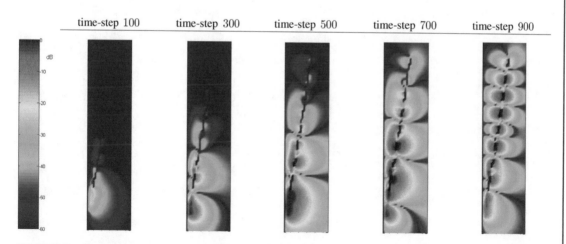

FIGURE 3.11: Snapshots of the normalized E_x component at the $y = 0$ plane for a semiconical substrate antenna with 10 microstrips and $\theta_0 = 9.5°$, $\psi = 90°$, $t_s = 4$ cells, $t_d = 6$ cells.

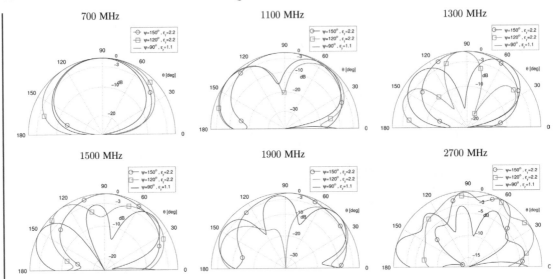

FIGURE 3.12: Radiation patterns of the E_q component for a semiconical substrate antenna with 10 microstrips, $\theta_0 = 7.5°$, and variable ψ, ε_r at diverse frequencies.

As with the CSA antenna, Figure 3.11 presents various snapshots of the E_x component (normalized to its maximum value) for a semiconical antenna of 10 microstrips with $\theta_0 = 9.5°$, $\psi = 90°$, $t_s = 4$ cells, and $t_d = 6$ cells. The results point out that both hybrid methods cope successfully with the multiple wave interactions in the vicinity of the cone, without inducing any nonphysical parasites.

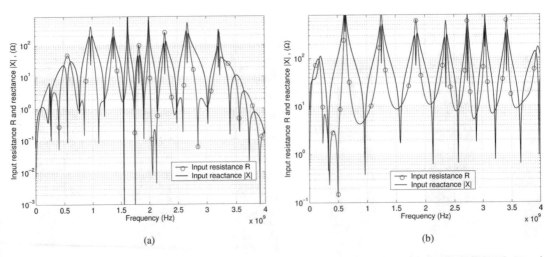

FIGURE 3.13: Input impedance of the semiconical substrate antenna with eight microstrips and (a) $\psi = 120°$, $\varepsilon_r = 2.2$ and (b) $\psi = 90°$ and $\varepsilon_r = 1.1$.

Moreover, Figure 3.12 gives the radiation patterns of the E_θ component for variable ψ and ε_r. From the plots, one can detect that an extended lobe with a maximum at 90° appears at low frequencies for every parameter set. However, as frequency increases, a number of supplementary lobes with almost the same power, are generated, First, at $\psi = 120°$ and then at $\psi = 90°$. Lastly, the input impedance (resistance R and reactance X) of a semiconical variant with eight microstrips and two sets of ψ and ε_r is depicted in Figure 3.13. Notice that for the case of $\psi = 120°$ and $\varepsilon_r = 2.2$, both quantities vary rather abruptly due to the resonances in the body of the antenna created by the dielectric. In contrast, these resonances become less and smoother for $\psi = 90°$ and $\varepsilon_r = 1.1$.

To complete our EMI analysis for microstrip-based antenna layouts, the contemporary class of log-periodic antennas (Figure 3.14) is examined. Such structures are normally used in wideband EMC measuring systems because they exhibit an independent behavior with regard to operating frequency [19, 22, 23]. In this work, two types are simulated, i.e., the *interleaved (IL)* and the *trapezoidal-toothed (TT)* antenna. Amid their design degrees of freedom, the electrically thin ground-plane-backed dielectric substrate is deemed very important. So, prospective bandwidth advancement may be accomplished by *increasing* the substrate thickness or *reducing* its dielectric constant. However, the former issue is regularly accompanied by inductive impedance offsets and the augmentation of the surface wave effect, especially in a strong EMI environment. Consequently, it becomes apparent that a meticulous process must be followed for the design of these antennas to avoid undue manufacturing costs.

The design parameters of a log-periodic antenna are α, β, R_{max}, R_{min}, the number of teeth N, the geometrical ratio τ, and the width ratio χ, as shown in Figure 3.14a. Specifically, τ and χ are given by

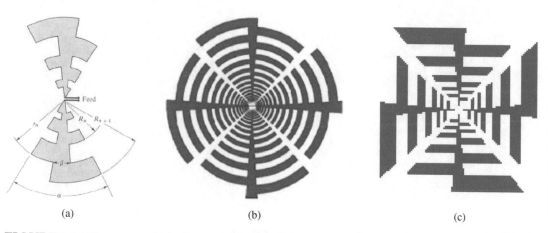

(a)   (b)   (c)

FIGURE 3.14: Geometry of (a) a log-periodic planar element, (b) an interleaved, and (c) a trapezoidal-toothed antenna.

$$\tau = \frac{R_n}{R_{n+1}} = \frac{r_{n-1}}{r_n}, \quad \chi = \frac{r_n}{R_{n+1}}, \tag{3.64}$$

and therefore, R_{n+1} and r_n are terms of geometrical progressions

$$R_{n+1} = \frac{R_{\min}}{\tau^{n+1}}, \quad r_n = \frac{r_{\min}}{\tau^n} \tag{3.65}$$

During the design process, τ and χ can be acquired from

$$\tau = \sqrt[N]{\frac{R_{\min}}{R_{\max}}}, \quad \chi = \sqrt{\tau} \tag{3.66}$$

The antennas of Figure 3.14b and 3.14c are realized through two metallic sheets, properly shaped and mounted on a dielectric board, whereas the feed sections are constructed by integrated tapered-microstrip baluns. Two out of four branches are located on the upper side of the board and the other two at its bottom.

The antennas, whose length is set to 90 mm, are simulated in terms of the FDTD and the MRTD methods for a computational domain of $150 \times 146 \times 50$ cells with $\Delta x = \Delta y = \Delta z = 0.833$ mm and $\Delta t = 1.458$ ps. Furthermore, termination of the unbounded space is attained by a six-cell PML absorber. The two ports are excited by hard sources, suitably phase-shifted to ensure circular polarization. Our interest principally focuses on the variation of α, N and the relative dielectric permittivity ε_r of the board. In this framework, Figure 3.15 shows the snapshots of the H_z component for the IL and the TT antenna, whereas Figure 3.16 gives their radiation patterns at the frequency of 2.12 GHz. From the results and taking into account the effect of EMI on the immunity behavior, it

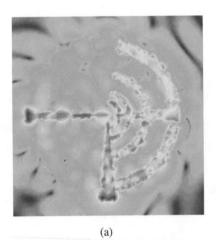

(a)

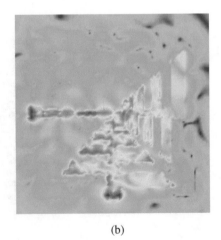

(b)

FIGURE 3.15: Snapshots of the H_z component for (a) an IL antenna with $N = 4$, $\alpha = 160°$, $\beta = 10°$, and (b) a TT antenna with $N = 5$, $\alpha = 80°$, $\beta = 10°$.

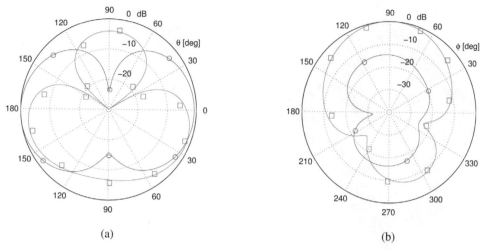

(a)   (b)

FIGURE 3.16: Radiation pattern at the x-plane for (a) an IL antenna with $N = 5$, $\alpha = 160°$, $\beta = 10°$, and (b) a TT antenna with $N = 5$, $\alpha = 80°$, $\beta = 10°$. The straight line refers to the E_φ and the circled one to the E_θ component, whereas the rectangles signify measurement data [23].

can be deduced that both time-domain techniques are in very close agreement with the measurement data [23], an issue that proves their suitability and accuracy for this type of EMC problems.

3.3.2   Antennas in Wireless Local Area Networks

The extensive utilization of wireless local area networks (WLANs) has activated intense research regarding the prediction of indoor wave propagation [24–31]. This is because the comprehension of the specific radio channel features is deemed critical for designing efficient communication systems. Nonetheless, in-building propagation is a highly complex process, because it occurs within environments possessing a variety of geometric and EMI issues. Owing to multiple interactions with walls, furniture, equipment, and people, received signals exhibit fading characteristics in space and spreading in time. Therefore, this paragraph is devoted to time-domain investigation of the prior setbacks in an effort to reveal the impact of electromagnetic radiation in the proper function of WLAN systems as parts of larger and more involved EMC arrangements.

    The indoor WLAN scene is the three-room office of Figure 3.17, which covers a space of 9.11 × 8.45 m [31]. The walls are 20 cm thick with a conductivity equal to $\sigma = 30$ mS/m and a relative dielectric constant of $\varepsilon_r = 5$. Rooms 2 and 3 are separated by a 10-cm-thick plastic partition of $\sigma = 0$ S/m and $\varepsilon_r = 3$. The same values are also used for the glass windows of rooms 1 and 2, whereas all metallic parts (window frames and supporting beams in the walls) have a conductivity

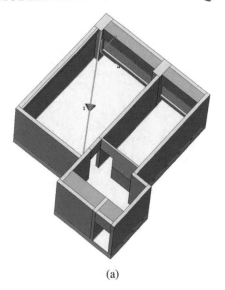

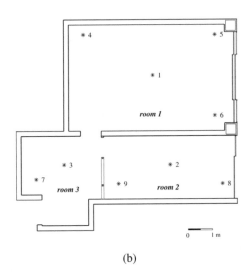

(a)   (b)

FIGURE 3.17: The investigated office region consisting of three rooms. (a) Perspective and (b) transverse cut with nine possible positions of the transmitter.

of $\sigma = 10^7$ S/m and $\varepsilon_r = 1$. All simulations, basically focusing on the frequencies of 2.44 and 5.80 GHz, are conducted by means of the FDTD method in a 1824×1710 square-cell lattice with $\Delta x = \Delta y = 5$ mm, terminated by a ten-cell PML. It is stressed that the transmitter at the nine preselected positions is modeled as an omnidirectional radiating source to achieve the best possible prediction outcomes for the system's channel behavior.

Nonetheless, before continuing with the numerical analysis, it is imperative to delve into some instructive WLAN aspects. Thus, to estimate the coverage capabilities of a wireless layout, time-harmonic variation is assigned to the transmitter's position, according to the operating frequency f_0. The electric field intensity is then computed by applying the discrete Fourier transform at f_0, after a sufficient amount of iterations for the attainment of steady state. In this way, the mean power can be locally derived by averaging the square of the normalized signal magnitude in a square area with 5λ sides. In particular, at point $(i\Delta x, j\Delta y)$, one gets

$$\langle P \rangle|_{i,j} = \frac{1}{N^2} \sum_{I=i-(N-1)/2}^{i+(N-1)/2} \sum_{J=j-(N-1)/2}^{j+(N-1)/2} E^2 \Big|_{I,J} \tag{3.67}$$

where N^2 is the number of samples. The effect of the averaging process is illustrated in Figure 3.18, along a path initiating from a transmitter at position 5λ with $f_0 = 2.44$ GHz. As observed, the received signal varies significantly for small displacements, yet small-scale fading is eliminated with the averaging process.

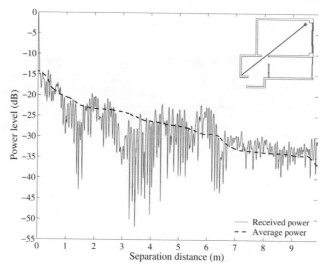

FIGURE 3.18: Received power levels toward the inlet figure path and spatially averaged values for the exclusion of small-scale fading.

Furthermore, for a statistical description of the received signal envelop at a local level, a common practice is to use Rician distributions. Rician fading is uniquely characterized by the K factor, representing the ratio of the power of the dominant component to that of the scattered ones. The first approach to calculate Rician parameters is in terms of the Rice distribution that best matches (in a least-squares sense) the actual distribution of the signal magnitude, whereas the second approach is based on the numerical solution of a specific equation. Because a single transmitted pulse reaches the receiver through different trajectories, it is regarded as a sequence of pulses. This phenomenon describes the time dispersion of the wireless channel, which may lead to EMI or intersymbol interference and degrade the WLAN performance [31]. The degree of dispersion is quantified via the recorded Power Density Probability, $p(t)$, from which the root mean square (rms) delay spread is obtained as

$$\sigma_\tau^2 = \frac{\int\limits_0^{t_{max}} (t - \bar{\tau})^2 p(t)\mathrm{d}t}{\int\limits_0^{t_{max}} p(t)\mathrm{d}t} \quad \text{with} \quad \bar{\tau} = \frac{\int\limits_0^{t_{max}} t p(t)\mathrm{d}t}{\int\limits_0^{t_{max}} p(t)\mathrm{d}t} \tag{3.68}$$

the Power Density Probability mean excess delay. Note that in all immunity simulations, these parameters are evaluated over a 150 MHz bandwidth around the operating frequencies.

Two of the factors that mainly affect the coverage characteristics of a WLAN are the operating frequency and the location of the access point. Table 3.1 provides a set of coverage statistics for

TABLE 3.1: Power Coverage (%) at diverse frequencies with regard to the area corresponding to the three rooms

POWER THRESHOLD (dB)	POSITION 1		POSITION 3		POSITION 5		POSITION 8		POSITION 9	
	2.44 GHz	5.80 GHz	2.44 GHz	5.80 GHz	2.44 GHz	5.80 GHz	2.44 GHz	5.80 GHz	2.44 GHz	5.80 GHz
≥ -20	3.12	1.47	2.93	1.44	3.19	1.82	1.94	1.56	2.39	1.71
≥ -25	43.99	21.98	24.29	15.77	26.14	14.49	17.49	16.14	16.75	10.76
≥ -30	58.70	55.99	51.91	48.66	54.24	51.88	43.57	40.38	39.81	35.11
≥ -35	91.09	77.20	71.87	66.38	80.37	67.84	85.03	78.60	75.45	61.45
≥ -40	98.19	96.68	97.03	95.13	97.19	91.43	99.81	99.52	96.29	96.23

transmitter positions 1, 3, 5, 8, and 9 at 2.44 and 5.80 GHz. The calculated values correspond to the area percentage within which the mean power levels satisfy specific thresholds (at 5-dB steps).

As expected, higher frequencies are more vulnerable to EMI attenuation [31]. In this context, Figures 3.19 and 3.20 illustrate the magnitude of the normalized electric field intensity and the power coverage maps, respectively, for different transmitter positions.

Inspecting the results from all transmitter positions, one may observe that the average power in rooms containing the source ranges from 24.23 to 21.41 dB at 2.44 GHz and from 25.39 to 22.54 dB at 5.80 GHz. These are the cases that satisfy line-of-sight (LoS) conditions. When the transmitter is located either in room 1 or 2, the differences between these two areas can be quite significant, because of the shadowing effect of the thick wall. At 2.44 GHz, the deviation of their mean power levels is between 6.75 and 11.19 dB. Similar remarks hold at 5.80 GHz as well.

Thus, when the antenna is in room 3, the path-loss in room 2 is lower than that in room 1, as a result of the weak signal obstruction caused by the partition. Moreover, Figure 3.21 depicts the electric field intensity at the above frequencies toward certain propagation paths, which intersect different rooms, thus revealing the variation of the WLAN signals in the presence of strong EMI.

Concerning LoS cases, the range of the average rms delay spread is within 11.07–15.02 ns at 2.44 GHz and 9.89–12.92 ns at 5.80 GHz. Figure 3.22 plots the rms delay spread along a path in rooms 2 and 3, when the source is at position 4 and operates at 5.35 GHz. Lower σ_τ values are found at points in LoS with the transmitter, determined by the width of the door between rooms 1 and 3 (dashed lines). Evidently, the rest of the line's points exhibit a more severe dispersion, especially those in room 3.

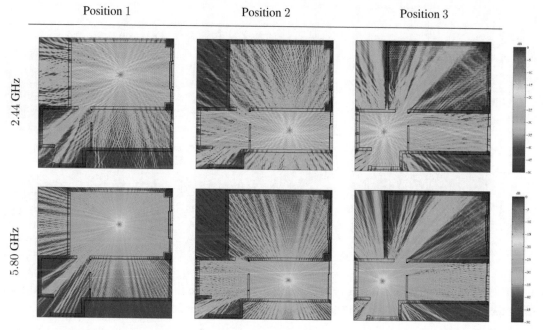

FIGURE 3.19: Magnitude of the electric field intensity (normalized to its maximum value) at 2.44 and 5.80 GHz for various positions of the transmitter.

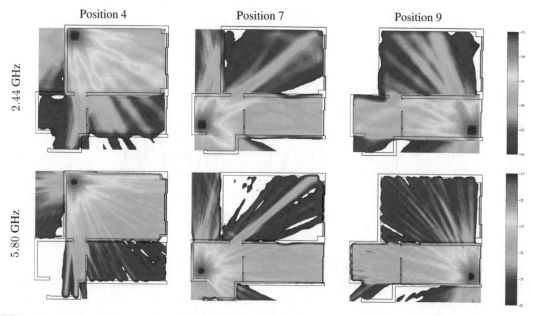

FIGURE 3.20: Power coverage maps at 2.44 and 5.80 GHz for different positions of the transmitter.

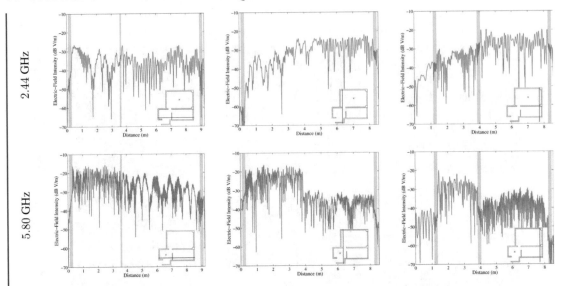

FIGURE 3.21: Electric field intensity level along the path shown in the inlet figure at 2.44 and 5.80 GHz for diverse positions of the transmitter. Gray areas indicate the location of concrete walls.

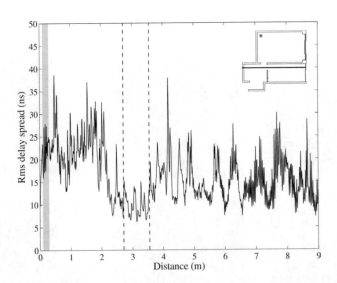

FIGURE 3.22: The rms delay spread along the path of the inlet figure. The transmitter is located at position 4 and operates at 5.80 GHz. The dashed lines indicate the door's opening, and the gray area corresponds to the wall.

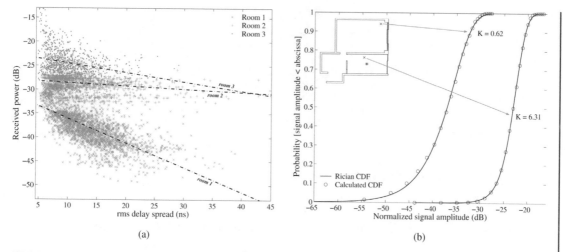

FIGURE 3.23: (a) Scatter plot of the power levels versus the rms delay spread in the three rooms for a transmitter located at position 7. (b) Numerical and fitted Rician of signal amplitude at two points for a transmitter located at position 2. Both simulations are conducted at 5.80 GHz.

Next, the dependence of the received power level on the rms delay spread is examined in the scatter plot of Figure 3.23a. The transmitter radiates from position 7 at 5.80 GHz. Higher power levels are associated with low values of the rms delay spread, again indicating weaker time dispersion for LoS regions (room 3). In addition, Figure 3.23b presents the numerical and fitted Rician factors, with the source being at position 2 at 5.80 GHz. For the two observation points marked in the inlet figure, the small-scale statistics are adequately described by means of Rician distributions, as verified by the good agreement with the computed distributions. Specifically, a high value of the Rician factor is calculated in the proximity of the transmitter, whereas K = 0.62 is obtained for the point in room 1. This indicates the absence of a strong path and, therefore, the statistical properties of the corresponding signal fluctuations resemble those of Rayleigh fading.

3.3.3 Specialized Configurations

In analyzing more sophisticated EMC radiating structures from an EMI perspective, first let us consider the single-element antenna of Figure 3.24, formed by a thin wire, with N turns, helically wound around a hemiellipsoidal surface with axis a along x and y directions and b along z direction. The antenna evolves above a finite $L_1 \times L_2$ PEC plate, whereas its helix is connected, via a short straight wire of length h, with a coaxial line on the ground plane. The single arm of the structure is described in cylindrical coordinates by

$$\rho = a \sin \theta \quad \text{and} \quad z = b \cos \theta, \tag{3.69}$$

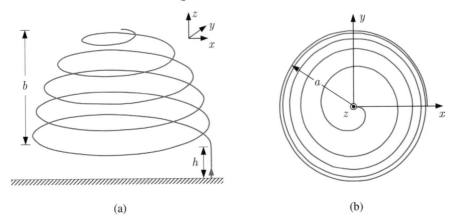

(a)   (b)

FIGURE 3.24: (a) Frontal view and (b) top view of the hemiellipsoidal helical antenna.

$$\theta = \cos^{-1}\left[\pm\left(\frac{\varphi}{2\pi N} - 1\right)\right] \quad \text{with} \quad 2\pi N \leq \varphi \leq 4\pi N, \tag{3.70}$$

where φ is measured counterclockwise from the $+x$ axis, and the $\pm$ sign indicates if the helix is right- or left-handed. When $b = a$, the hemispherical helical counterpart is obtained, which exhibits desirable radiation characteristics, such as low axial ratio and slightly fluctuating input impedance. Typical dimensions for our simulations are: $a = 1.9615$ cm, $b = 1.9615, 1.4708, 0.9802$ cm, and $h = 0.4906$ cm, with the ground plane chosen to be a square plate of 19.992 cm. For the excitation,

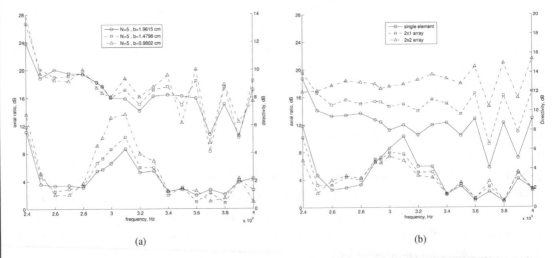

(a)   (b)

FIGURE 3.25: Axial ratio and directivity versus frequency for (a) $N = 5$ as well as three different values of b and (b) for a single-element, 2×1 and 2×2 array consisting of hemiellipsoidal helical antennas.

a modulated Gaussian pulse, centered at 2.94 GHz, with a 6-GHz bandwidth is used. Using the FDTD method, the domain is discretized through a $\lambda_0/104$ resolution (λ_0 is wavelength at the central frequency), yielding a $224 \times 224 \times 40$ mesh. Figure 3.25a shows the axial ratio and power gain for three b values over the range of 2.4–4.0 GHz. Note that axial ratio *does not* seriously vary at 2.5–2.8 GHz, remaining lower than 3 dB. Similar behavior is acquired for the power at 2.5–3.4 GHz, where the difference between its maximum and minimum is also less than 3 dB.

Keeping the rest of the parameters the same, a 2×1 and 2×2 array, whose elements are hemiellipsoidal antennas, is examined. All elements are excited in phase so as to obtain a maximum in the radiation pattern for $\theta = 0°$. Figure 3.25b shows the axial ratio and directivity of the prior arrays (element distances dx = dy = 5.10 cm) and compares their performance with the single-element one. Results indicate that directivity increases 2 dB when elements are doubled, whereas antenna polarization does not greatly change.

The subsequent application is the modern fractal volume antenna of Figure 3.26 that consists of two intersecting Sierpinski gaskets of equal height [32–34]. In fact, the use of fractal concepts in the design of EMC radiating structures has received remarkable attention because their geometry can be efficiently implemented when certain features, such as miniaturization or multiband behavior, are required. The specific antenna resonates at various frequencies, owing to the analogous character of its gaskets, and therefore it is likely to suffer from the consequences of EMI. To model this radiator, the conformal FDTD algorithm has been selected. Thus, by properly manipulating the exact location of every field component near the antenna, the technique establishes an optimized local mesh deformation to effectively model arbitrary curves. Both primary and dual grids are

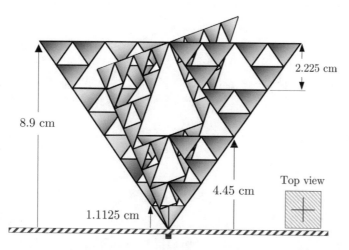

FIGURE 3.26: General geometry of the Sierpinski double-gasket antenna with arbitrary feeding flare angles.

consistently constructed via a rigorous projection operator, whereas the rest of the domain receives the usual FDTD discretization. In this sense, Ampère's law, for $u = x, y, z$, is given by

$$E_u\big|_{i,j,k}^{n+1} = q_u^E E_u\big|_{i,j,k}^{n} + p_u^E \Delta t \sum_{v \neq u} s_v w_v^H \left(l_v^H, A_v^H \right) H_v\big|_{i,j,k}^{n+1/2}, \tag{3.71}$$

where the summation involves all $H_v(v \neq u)$ magnetic components (toward $s_v = \pm 1$) along the contour (of area A_v) formed by the corresponding dual edges (of length l_v) that enclose the primary electric field E_u. Also, w_v are coefficients depending on l_v, A_v and q_u, p_u coefficients specified by material properties. Superscripts E or H denote the field type that contributes to each parameter. Similarly, Faraday's law becomes

$$H_u\big|_{i,j,k}^{n+1/2} = q_u^H H_u\big|_{i,j,k}^{n-1/2} - p_u^H \Delta t \sum_{v \neq u} s_v w_v^E \left(l_v^E, A_v^E \right) E_v\big|_{i,j,k}^{n}, \tag{3.72}$$

in which duality with (3.72) is obvious. Here, the monopole is made of two gaskets with 90° flare angles. The total height of the model is 8.9 cm and both of its individual elements are constructed after four iterations with a scaling factor of 2. The entire arrangement, resonating at 0.5 and 1.5 GHz, is mounted over an infinite ground plane and excited by a resistive voltage source with an input resistance of 50 Ω.

Figure 3.27 illustrates the X–Y and X–Z cuts of the computed radiation patterns with a 40-dB dynamic range. Apparently, the outcomes unveil a noteworthy degree of similarity, while the number of lobes is kept constant throughout the resonant frequencies. All discretizations involved a 100 × 100

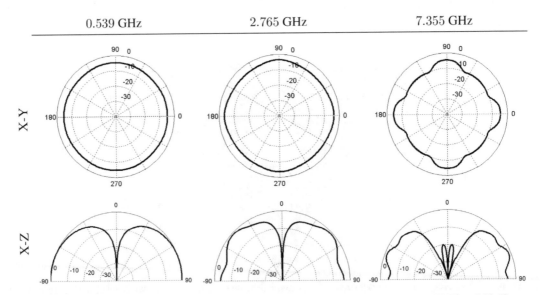

FIGURE 3.27: Radiation patterns of the Sierpinski double-gasket antenna at the X–Y and X–Z cuts.

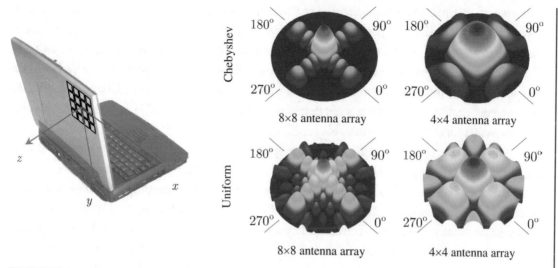

FIGURE 3.28: Geometry and radiation patterns of a WLAN PCB antenna array mounted on a laptop PC for various numbers of elements and excitation distributions.

$\times$ 51 conformal FDTD grid (about 70% smaller than the staircase lattice) using a temporal increment $\Delta t = 4.817$ ps.

As a final example, assume the PCB antenna array of Figure 3.28, mounted on a laptop PC for wireless communication [19]. The radiator is placed on the cover of the computer and has a 4×4 or 8×8 layout. Moreover, its elements are excited by current densities, whose phases follow a Chebyshev or a uniform distribution. The structure is analyzed through the FIT and the FETD method, with the infinite boundaries backed by an eight-cell PML. Figure 3.28 displays the radiation patterns of the aforesaid configurations and clearly indicates the efficiency of the selected time-domain techniques for the solution of this demanding problem.

3.4   SHIELDING APPLICATIONS AND SIGNAL INTEGRITY

To guarantee the unobstructed operation of a device in an undesirable EMI environment and increase its immunity capabilities, the solution of shielding could, indeed, be proven very helpful. In the field of electromagnetism, shielding may mean different things in different circumstances and therefore it is viewed as being either electric or magnetic in nature. Alternatively, it can be considered as arising from eddy currents induced on the shield material, which in practice is the only mechanism active when counteracting EMI. On the other hand, many real-world applications require the use of a volumetric closed structure, commonly approximated by a spherical or cylindrical shape [37–45]. It is, however, foremost to emphasize that a correct choice of shields and other elements of an EMI protection scheme must take into account *not only* the surroundings and its properties, but also the user preferences and mission of the EMC system. Obviously, the disadvantages in any shield are the openings on its surface, which in everyday equipment are inevitable. Generally, these penetrations provide the major degradations in the total performance and spoil the signal integrity of the protected component. As a natural consequence, the verification of a shield's competences should be a very systematic task with a lot of "trial and error" experiments. To this aim, time-domain modeling can be a valuable certification instrument.

Based on the preceding remarks, the first application of this section is the rectangular shielding enclosure in Figure 3.29a with a partitioning sheet in its interior. The latter shield has an elliptical aperture that is intended to test the aptitude of the entire cavity in the proper isolation of a device placed in its right-hand part. An ordinary set of dimensions is: $l_x = 50$ cm, $l_y = 110$ cm, $l_z = 60$ cm, $l = 45$ cm, whereas the excitation is launched by a vertical coaxially fed monopole. For the rigorous solution of this relatively difficult problem, the nonstandard (ND) FDTD method (see Chapter 5) is implemented, thus leading to the coarse mesh of $44 \times 56 \times 22$ cells with $\Delta t = 25.131$ ps. Figure 3.29b gives the shielding efficiency of the cavity computed at an observation point in front of the elliptical aperture. For the sake of comparison, two conventional—and much more refined (almost 80%)—FDTD realizations along with the reference solution of Ref. [41] are also included. Clearly, the selected technique provides the most rigorous results, overwhelming the simple FDTD ones, especially in the evaluation of the three frequency peaks attributed to the presence of the aperture.

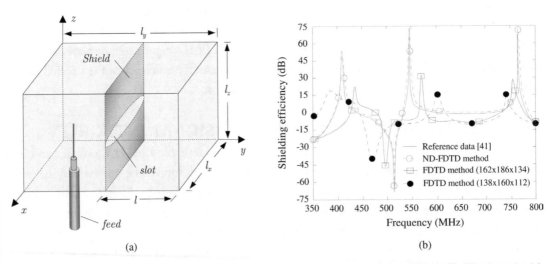

(a)   (b)

FIGURE 3.29: (a) Geometry of a rectangular shielding cavity with an internal partitioning shield. (b) Shielding efficiency of the structure.

To determine the influence of a correctly constructed shielding in the signal integrity of a device, let us presume the three-port, inclined-slot coupled cylindrical cavity of Figure 3.30a.

The dimensions of our shielding configuration are: $a_1 = 10.72$ mm, $a_2 = 26.43$ mm, $a_3 = 38.24$ mm, $b_1 = 8.39$ mm, $b_2 = 11.08$ mm, and $b_3 = 28.15$ mm. Because of its curved parts, the structure is analyzed by means of the hybrid FDTD/ FVTD and the FETD method, which produce very accurate and convergent computational models. In this manner, the domain is discretized into $82 \times 104 \times 176$ cells or 560,284 elements with $\Delta t = 2.845$ ps. Also, the slot inclination angle is $\theta = 45°$, while the open ends are truncated by an eight-layer PML. Figure 3.30b shows the magnitude of different S parameters (excitation is at port 1) compared with those obtained by the modal method of Rajami et al. [45]. As discerned, both techniques are in very good agreement with the reference data, requiring significantly lower resources (practically 60%) than the usual second-order FDTD approach.

The third EMC arrangement examined for its signal integrity is a conducting aperture with two rectangular waveguides coupled to its surface, depicted in Figure 3.31a. All simulations are conducted in terms of an enhanced alternating-direction implicit (ADI)-FDTD method, based on a family of multidirectional spatial/temporal operators (see Chapter 5) that minimize dispersion errors as time-steps *surpass* the Courant condition. A standard set of dimensions are: $l_1 = 10.84$ mm, $l_2 = 29.16$ mm, $l = 40.92$ mm, $d_1 = 4.05$ mm, $d_2 = 1.43$ mm, and $s = 1.57$ mm, with the computational

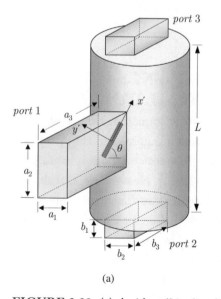

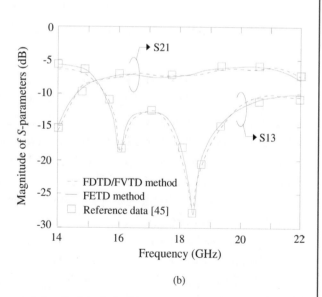

(a)   (b)

FIGURE 3.30: (a) A sidewall inclined-slot coupled cylindrical shielding cavity and (b) magnitude of diverse S parameters.

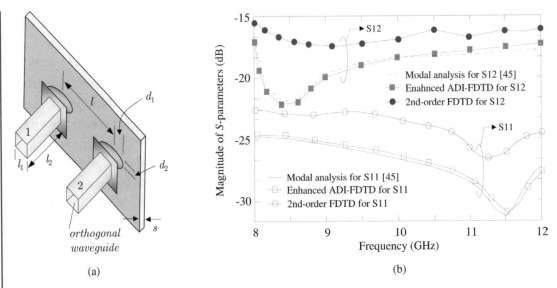

FIGURE 3.31: (a) An aperture with two elliptical irises connected to two rectangular waveguides and (b) magnitude of the S parameters.

domain including $24 \times 88 \times 56$ cells. Figure 3.31b presents the S parameters of the aperture and compares the outcomes with the findings of the second-order FDTD technique. Note the promising coincidence of the reference data [45] with the ADI-FDTD plots for a Δt selected 24 times beyond the stability limit as well as the failure of the FDTD algorithm to supply satisfactory values.

3.5   EMI INVESTIGATION OF RADIO FREQUENCY MICROELECTROMECHANICAL SYSTEM LAYOUTS

As already noted in the previous chapter, radio frequency microelectromechanical systems (RF MEMS) constitute a thriving field of modern electromagnetics with a large assortment of applications. Combining the small size with the high-end capabilities, these structures can substitute for traditional devices, such as filters, switches, or power dividers [46–51]. Nonetheless, and despite their improved behavior, in the case of a tense EMI exposure they may lose these merits due to unpredictable wave interactions. So, a careful investigation is required before their actual fabrication. For this goal, some indicative time-domain simulations are included in this section.

The first problem is the two-port coaxial waveguide of Figure 3.32, which involves a set of variable RF MEMS actuators for selective mode propagation. By changing the inclination angle of the latter, the whole device can exclude the transmission of certain wave constituents and hence act as an adjustable bandpass filter.

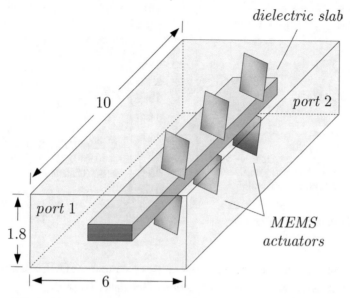

FIGURE 3.32: An RF two-port coaxial waveguide with two rows of MEMS actuators. All dimensions are in mm.

TABLE 3.2: Resonances of the RF MEMS-based two-port waveguide						
REFERENCE (GHz) [46]	METHOD (CFLN)	COMPUTED (GHz)	ERROR (%)	LATTICE (CELLS)	CPU TIME	MAXIMUM DISPERSION
Two actuators 12.832	FDTD	11.884	7.38	$168 \times 186 \times 72$	13.1 h	1.58742×10^{-1}
	ADI-FDTD (29)	12.831	0.01	$46 \times 54 \times 32$	55 min	2.12378×10^{-11}
Four actuators 15.921	FDTD	14.604	8.27	$210 \times 242 \times 84$	12.3 h	3.72197
	ADI-FDTD (34)	15.914	0.04	$62 \times 68 \times 36$	47 min	3.64103×10^{-10}
Six actuators 17.642	FDTD	15.774	10.59	$238 \times 274 \times 96$	11.8 h	6.59104
	ADI-FDTD (37)	17.631	0.06	$70 \times 78 \times 46$	41 min	4.37025×10^{-10}

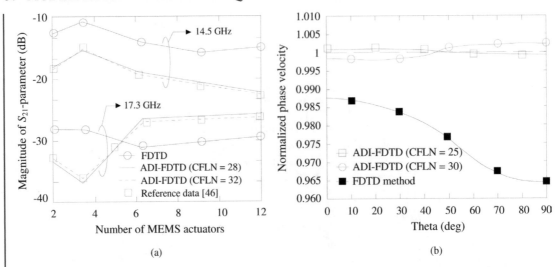

(a)

(b)

FIGURE 3.33: (a) Magnitude of the S_{21} parameter versus the number of MEMS actuators in the wave-guide, and (b) normalized phase velocity as a function of incidence angle θ.

The waveguide is modeled via the advanced ADI-FDTD technique, previously used. Table 3.2 summarizes the resonance frequencies for three sets of actuators along with several interest-ing realization aspects. Note that CFLN stands for the Courant-Friedrich-Levy number (CFLN), namely, the ratio between the temporal increment of the ADI-FDTD algorithm and the one dic-tated by the usual FDTD stability criterion. Moreover, the variation of the important S_{21} parameter

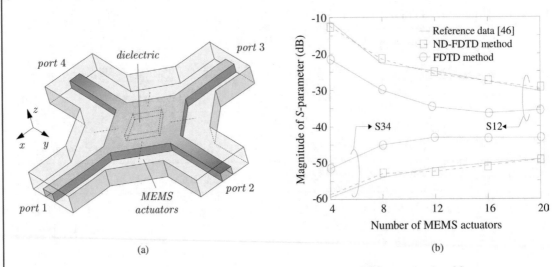

(a)

(b)

FIGURE 3.34: A four-port RF MEMS splitter. (a) Geometry and (b) magnitude of S parameters.

versus the amount of MEMS is presented in Figure 3.33a. Apparently, the selected ADI-FDTD schemes—compared to the second-order FDTD approach—are notably accurate and economical, achieving drastically diminished dispersion errors and consistent phase velocities, even for large CFLNs, as also derived from Figure 3.33b.

As a last realistic example, let us assume the four-port RF MEMS splitter of Figure 3.34a comprising a number of actuators for the control of modal interactions. The actuators are located in a $2.2 \times 0.6 \times 0.4$ mm region, whereas the outer conductor's cross section is 6.2×5.8 mm and that of the inner conductor is 1.2×0.8 mm.

Figure 3.34b illustrates the variation of two S-parameters as a function of MEMS actuators, evaluated by means of the ND-FDTD method and a six-cell PML. From the results, the superiority of the improved technique is easily discernible, particularly when compared with the conventional FDTD algorithm and the reference data [47].

REFERENCES

1. T. Weiland, "A discretization method for the solution of Maxwell's equations for six-component fields," *Electron. Commun. (AEÜ)*, vol. 31, no. 3, pp. 116–120, 1977.

2. T. Weiland, "Time domain electromagnetic field computation with finite difference methods," *Int. J. Numer. Model.: Electron. Netw., Device, Fields*, vol. 3, pp. 295–319, 1996.

3. J.-F. Lee, R. Lee, and A. C. Cangellaris, "Time-domain finite-element methods," *IEEE Trans. Antennas Propagat.*, vol. 45, no. 3, pp. 430–442, Mar. 1997.

4. D. Jiao and J.-M. Jin, "A general approach for the stability analysis of the time-domain finite-element method for electromagnetic simulations," *IEEE Trans. Antennas Propagat.*, vol. 50, no. 11, p. 1624–1632, Nov. 2003.

5. M. Feliziani and F. Maradei, "Point matched finite element-time domain method using vector elements," *IEEE Trans. Magn.*, vol. 30, no. 5, pp. 3184–3187, Sept. 1994. doi:10.1109/20.312614

6. R. Holland, V. P. Cable, and L. C. Wilson, "Finite-volume time-domain (FVTD) techniques for EM scattering," *IEEE Trans. Electromagn. Compat.*, vol. 33, no. 4, pp. 281–294, Nov. 1991. doi:10.1109/15.99109

7. C. Fumeaux, D. Baumann, and R. Vahldieck, "A generalized local time-step scheme for efficient FVTD simulations in strongly inhomogeneous meshes," *IEEE Trans. Microwave Theory Tech.*, vol. 52, no. 3, pp. 1067–1076, Mar. 2004. doi:10.1109/TMTT.2004.823595

8. M. Fujii and W. R. J. Hoefer, "A three-dimensional Haar wavelet-based multiresolution analysis similar to the FDTD method—Derivation and application," *IEEE Trans. Microwave Theory Tech.*, vol. 46, no. 12, pp. 2463–2475, Dec. 1998.

9. M. Krumpholz and L. P. B. Katehi, "MRTD: New time-domain schemes based on multiresolution analysis," *IEEE Trans. Microwave Theory Tech.*, vol. 44, no. 4, pp. 555–571, Apr. 1999.

10. N. Bushyager and M. Tentzeris, *MRTD (Multi Resolution Time Domain) Method in Electromagnetics*. San Rafael, CA: Morgan & Claypool Publishers, 2005. doi:10.2200/S00009ED1V 01Y200508CEM002

11. Special issue on "High-power electromagnetics (HPEM) and international electromagnetic interference (IEMI)," W. A. Radasky, C. E. Baum, and M. W. Wik, Eds., *IEEE Trans. Electromagn. Compat.*, vol. 46, no. 3, Aug. 2004, pp. 314–439. doi:10.1109/TEMC.2004.831899

12. Y. Matsumoto, M. Takeuchi, K. Fujii, A. Sugiura, and Y. Yamanaka, "Performance analysis of interference problems involving DS-SS WLAN systems and microwave ovens," *IEEE Trans. Electromagn. Compat.*, vol. 47, no. 1, pp. 45–53, Feb. 2005. doi:10.1109/TEMC.2004.842114

13. H. Bağci, A. E. Yilmaz, J.-M. Jin, and E. Michielssen, "Fast and rigorous analysis of EMC/EMI phenomena on electrically large and complex cable-loaded structures," *IEEE Trans. Electromagn. Compat.*, vol. 49, no. 2, pp. 361–381, Nov. 2005.

14. S. Iskra, B. W. Thomas, R. McKenzie, and J. Rowley, "Evaluation of potential GPRS 900/1800-MHz and WDCMA 1900-MHz interference to consumer electronics," *IEEE Trans. Electromagn. Compat.*, vol. 47, no. 4, pp. 951–962, Nov. 2005. doi:10.1109/TEMC.2005.857363

15. F. Fiori, "Design of an operational amplifier input stage immune to EMI," *IEEE Trans. Electromagn. Compat.*, vol. 49, no. 4, pp. 834–839, Nov. 2007.

16. J. Cago, J. Balcells, D. Conzález, M. Lamich, J. Mon, and A. Santolaria, "EMI susceptibility model of signal conditioning circuits based on operational amplifiers," *IEEE Trans. Electromagn. Compat.*, vol. 49, no. 4, pp. 849–859, Nov. 2007.

17. J. Qin, O. M. Ramahi, and V. Granatstein, "Novel planar electromagnetic bandgap structures for mitigation of switching noise and EMI reduction in high-speed circuits," *IEEE Trans. Electromagn. Compat.*, vol. 49, no. 3, pp. 661–669, Aug. 2007.

18. K. L. Wong, *Compact and Broadband Microstrip Antennas*. Piscataway, NJ: Wiley Interscience, 2002. doi:10.1002/0471221112.ch3

19. C. A. Balanis, *Antenna Theory: Analysis and Design*, 3rd ed. New York: IEEE Press and Wiley Interscience, 2005.

20. T. H. Hertel and G. S. Smith, "Analysis and design of two-arm conical spiral antennas," *IEEE Trans. Electromagn. Compat.*, vol. 44, no. 1, pp. 25–37, Feb. 2002. doi:10.1109/15.990708

21. T. H. Hertel and G. S. Smith, "On the dispersive properties of the conical spiral and its use for pulsed radiation," *IEEE Trans. Antennas Propagat.*, vol. 51, no. 7, pp. 1426–1433, Jul. 2003. doi:10.1109/TAP.2003.813602

22. A. Gadzina and P. Slobodzian, "A compact dual-port, dual-band planar microstrip antenna," *Microwave Opt. Technol. Lett.*, vol. 34, pp. 302–305, Aug. 2002. doi:10.1002/mop.10443

23. K. M. P. Aghdam, R. Faraji-Dana, and J. Rashed-Mohassel, "Compact dual-polarisation planar log-periodic antennas with integrated feed circuit," *IEEE Microwave Antennas Propagat.*, vol. 152, pp 359–366, 2005. doi:10.1049/ip-map:20050091

24. I. Cuinas and M. G. Sanchez, "Measuring, modeling, and characterizing of indoor radio channel at 5.8 GHz," *IEEE Trans. Veh. Technol.*, vol. 50, pp. 526–535, Mar. 2001.

25. V. Degli-Esposti, G. Lombardi, C. Passerini, and G. Riva, "Wide-band measurement and ray-tracing simulation of the 1900-MHz indoor propagation channel: Comparison criteria and results," *IEEE Trans. Antennas Propagat.*, vol. 49, pp. 1101–1110, July 2001. doi:10.1109/8.933490

26. J. Kivinen, X. Zhao, and P. Vainikainen, "Empirical characterization of wideband indoor radio channel at 5.3 GHz," *IEEE Trans. Antennas Propagat.*, vol. 49, pp. 1192–1203, Aug. 2001. doi:10.1109/8.943314

27. N. J. McEwan, R. A. Abd-Alhameed, E. M. Ibrahim, P. S. Excell, and N. T. Ali, "Compact WLAN disc antennas," *IEEE Trans. Antennas Propagat.*, vol. 50, pp. 1862–1864, Dec. 2002. doi:10.1109/TAP.2002.807368

28. T. K. Sarkar, Z. Ji, K. Kim, A. Medouri, and M. Salazar-Palma, "A survey of various propagation models for mobile communication," *IEEE Antennas Propagat. Mag.*, vol. 45, pp. 51–82, June 2003. doi:10.1109/MAP.2003.1232163

29. Z. Yun, M. F. Iskander, and Z. Zhang, "Complex-wall effect on propagation characteristics and MIMO capacities for an indoor wireless communication environment," *IEEE Trans. Antennas Propagat.*, vol. 52, pp. 914–922, Apr. 2004. doi:10.1109/TAP.2004.825691

30. Y.-T. Liu, C.-W. Su, K.-L. Wong, and H.-T. Chen, "An air-substrate narrow-patch microstrip antenna with high radiation performance for 2.4 GHz WLAN access point," *Microwave Opt. Technol. Lett.*, vol. 43, pp. 189–192, Nov. 2004. doi:10.1002/mop.20416

31. T. T. Zygiridis, E. P. Kosmidou, K. P. Prokopidis, N. V. Kantartzis, C. S. Antonopoulos, K. I. Petras, and T. D. Tsiboukis, "Numerical modeling of an indoor wireless environment for the performance evaluation of WLAN systems," *IEEE Trans. Magn.*, vol. 42, no. 4, pp. 839–842, Apr. 2006.

32. N. V. Kantartzis, T. T. Zygiridis, and T. D. Tsiboukis, "A nonstandard higher-order FDTD algorithm for 3-D arbitrarily- and fractal-shaped antenna structures on general curvilinear lattices," *IEEE Trans. Magn.*, vol. 38, no. 2, pp. 737–740, Mar. 2002. doi:10.1109/20.996191

33. T. T. Zygiridis, N. V. Kantartzis, and T. D. Tsiboukis, "A novel Sierpinski double-gasket antenna investigated with a 3-D FDTD conformal technique," *IEEE Electron. Lett.*, vol. 38, no. 3, pp. 107–109, 2002.

34. T. T. Zygiridis, N. V. Kantartzis, T. V. Yioultsis, and T. D. Tsiboukis, "Higher-order approaches of FDTD and TVFE methods for the accurate analysis of fractal antenna arrays," *IEEE Trans. Magn.*, vol. 39, no. 2, pp. 1230–1233, Mar. 2003. doi:10.1109/TMAG.2003.810204

35. S. V. Georgakopoulos, C. R. Birtcher, and C. A. Balanis, "Coupling modeling and reduction techniques of cavity-backed slot antennas: FDTD versus measurements," *IEEE Trans. Electromagn. Compat.*, vol. 43, no. 3, pp. 261–272, Aug. 2001. doi:10.1109/15.942599

36. K. P. Prokopidis and T. D. Tsiboukis, "FDTD algorithm for microstrip antenna with lossy substrate using higher-order schemes," *Electromagnetics*, vol. 2, no. 5, pp. 301–315, 2004. doi:10.1080/02726340490457764

37. M. A. Gkatzianas, G. I. Ballas, C. A. Balanis, C. R. Birtcher, and T. D. Tsibukis, "Thin-slot/thin-layer subcell FDTD algorithms for EM penetration through apertures," *Electromagnetics*, vol. 23, no. 2, pp. 119–133, 2003. doi:10.1080/02726340390159469

38. E. S. Siah, K. Sertel, J. L. Volakis, V. V. Liepa, and R. Wiese, "Coupling studies and shielding techniques for electromagnetic penetration through apertures on complex cavities and vehicular platforms," *IEEE Trans. Electromagn. Compat.*, vol. 45, no. 2, pp. 245–257, May 2003.

39. M. Isteni and R. G. Olsen, "A simple hybrid method for ELF shielding by imperfect finite planar shields," *IEEE Trans. Electromagn. Compat.*, vol. 46, no. 4, pp. 199–207, May 2004. doi:10.1109/TEMC.2004.826888

40. M. S. Sarto, "Electromagnetic shielding of thermoformed lightweight plastic screens," *IEEE Trans. Electromagn. Compat.*, vol. 46, no. 4, pp. 588–596, Nov. 2004. doi:10.1109/TEMC.2004.837843

41. L. Klinkenbusch, "On the shielding effectiveness of enclosures," *IEEE Trans. Electromagn. Compat.*, vol. 47, no. 3, pp. 589–601, Aug. 2005. doi:10.1109/TEMC.2005.853162

42. N. V. Kantartzis, T. D. Tsiboukis, and E. E. Kriezis, "A topologically consistent class of 3-D higher-order curvilinear FDTD schemes for dispersion-optimized EMC and material modeling," *J. Mater. Process. Technol.*, vol. 161, pp. 210–217, 2005. doi:10.1016/j.jmatprotec.2004.07.027

43. T. Konefal, J. F. Dawson, A. C. Marvin, M. P. Robinson, and S. J. Porter, "A fast circuit model description of the shielding effectiveness of a box with imperfect gaskets or apertures covered by thin resistive sheets," *IEEE Trans. Electromagn. Compat.*, vol. 48, no. 1, pp. 134–144, Feb 2006. doi:10.1109/TEMC.2006.870703

44. M. S. Sarto and A. Tamburrano, "Innovative test method for the shielding effectiveness measurement of conductive thin films in a wide frequency range," *IEEE Trans. Electromagn. Compat.*, vol. 48, no. 2, pp. 331–341, May 2006. doi:10.1109/TEMC.2006.874664

45. V. Rajami, C. F. Bunting, M. D. Deshpande, and Z. A. Khan, "Validation of modal/MoM in shielding effectiveness studies of rectangular enclosures with apertures," *IEEE Trans. Electromagn. Compat.*, vol. 48, no. 4, pp. 348–353, May 2006.

46. M. Daneshmand, R. Mansour, and N. Sarkar, "RF MEMS waveguide switch," *IEEE Trans. Microwave Theory Tech.*, vol. 52, no. 12, pp. 2651–2657, Dec. 2004.

47. J. Schoebel, T. Buck et al., "Design considerations and technology assessment of phased-array systems RF MEMS automotive applications," *IEEE Trans. Microwave Theory Tech.*, vol. 53, no. 6, pp. 1968–1975, Jun. 2005. doi:10.1109/TMTT.2004.838269

48. D. Anagnostou, G. Zheng, M. Chryssomalis, J. Lyke, G. E. Ponchak, J. Papapolymerou, and C. G. Christodoulou, "Design, fabrication, and measurements of an RF-MEMS-based self-similar antenna," *IEEE Trans. Antennas Propagat.*, vol. 54, no. 2, pp. 422–431, Feb. 2006. doi:10.1109/TAP.2005.863399

49. E. Edril, K. Topalli, M. Unlu, O. A. Civi, and T. Akin, "Frequency tunable microstrip patch antenna using RF MEMS technology," *IEEE Trans. Antennas Propagat.*, vol. 55, no. 4, pp. 1193–1195, Apr. 2007.

50. Y. Munemasa, M. Mita, T. Takano, and M. Sano, "Lightwave antenna with a small aperture manufactured using MEMS processing technology," *IEEE Trans. Antennas Propagat.*, vol. 55, no. 11, pp. 3046–3051, Nov. 2007.

51. N. V. Kantartzis, T. I. Kosmanis, and T. D. Tsiboukis, "Fully nonorthogonal higher-order accurate FDTD schemes for the systematic development of 3-D reflectionless PMLs in general curvilinear coordinate systems," *IEEE Trans. Magn.*, vol. 36, no. 4, pp. 912–916, Jul. 2000. doi:10.1109/20.877591

• • • •

CHAPTER 4

Bioelectromagnetic Problems—Human Exposure to Electromagnetic Fields

4.1   INTRODUCTION

One of the most rapidly developing areas in EMC analysis is the category of bioelectromagnetic problems, whose correct characterization and accurate solution is the subject of a really impressive research. Being rather complicated and requiring the sound knowledge of numerous other scientific fields, these real-world applications are strongly connected to human health and therefore any attempt for their modeling should be very systematic and well established. In essence, since the beginning of the previous decade, consumers experienced a tremendous technological outgrowth with an emphasis to the sector of personal wireless communications. Nonetheless, when the first zest on these indisputable achievements subsided, the major apprehension of the community focused on the possible negative effects of the electromagnetic power absorbed by certain parts of the human body. Actually, with the constant market demands for minimal size and high-end capabilities, all features of wireless components and especially of cellular phones have been reengineered. Despite their enhanced quality and the manufacturers' confirmations, there is still a serious anxiety whether these devices do comply with the international safety standards for microwave exposure to the user.

Consequently, special attention must be drawn to the design and performance of the previous structures, without neglecting their excessive fabrication expenditures. Obviously, time-domain computational methods are a powerful predictive tool for this purpose, because they offer prompt and rigorous near- and far-field results for basically arbitrary setups of inhomogeneous media. Although several methods have been hitherto used, relevant literature substantiates that the FDTD algorithm dominates this class of problems [1–3]. So, in this chapter, a broad analysis (along with thermal simulations) on the hazards of electromagnetic radiation is conducted for various cellular phones and wireless local area networks (WLANs), aiming at antenna interactions with the involved tissues of the human body.

4.2 BASIC FEATURES OF TIME-DOMAIN NUMERICAL SIMULATIONS

Taking into account the decisive contribution of modern electromagnetic devices in everyday life quality and the constantly increasing concern for the consequences of human exposure to electromagnetic fields, it becomes obvious that numerical simulations can be proven a very instructive means toward the precise characterization of any possible hazardous effects. Furthermore, the relative extended frequency spectrum where such apparatuses operate indicates, even more prominently, the significance of time-domain methods in the analysis of these problems. Actually, the impressive evolution of computer systems in conjunction with the rapid growth of advanced numerical techniques, gave the opportunity for the discretization of very complex applications, such as the human head or body. Therefore, during the past decade, a voluminous assortment of studies, with an emphasis on radiation thermal effects, has been presented [4–47]. Based on the above remarks, the present section conducts a comprehensive investigation of electromagnetic field distribution in the human body, radiated by several types of cellular phone and WLAN antennas via a set of dispersive FDTD-oriented approaches. In particular, the comparative examination involves monopole, helical, side-mounted planar inverted-F antenna (PIFA), and patch components. The formulation is improved by subgridding schemes, which attain advanced accuracy in the modeling of thin geometries, whereas an efficient thermal model is introduced for the estimation of the corresponding effects. In this way, one can extract a notable amount of outcomes along with informative graphical representations of the *specific absorption rate (SAR)* in the interior of the human body [4–13], not feasible via experimental procedures. As will be deduced, the whole assessment reveals the occurrence of considerable differences among the antennas and the most striking: the high absorption from the head, which, aside from its possible biological influence, results in serious degradation of the antenna efficiency.

4.2.1 Computational Models of the Human Head and Body

4.2.1.1 The Case of Human Head. Being an extremely complicated structure, the human head requires a very large amount of volume elements (voxels) for its correct geometrical representation. Essentially, for a rational compromise between modeling precision and computational burden, the selection of a 3-cm cubic voxel seems to be sufficient. The discretization then uses diverse interpolation algorithms to the data of high-resolution magnetic resonance images (at consecutive distances of 4 mm) to produce robust lattices. The next step is the truncation of the infinite space via the pertinent ABC. Here, the conventional form of a powerful eight-layer PML (perfectly matched layer) absorber has enabled the reduction of the total domain to a $101 \times 107 \times 109$ mesh, thus yielding simulation times of a few hours' order. A transverse cut of the final model can be viewed in Figure 4.1 for the situation of electromagnetic field and thermal computations, respectively.

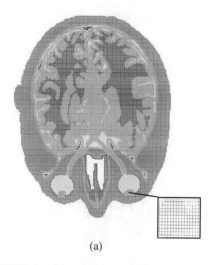
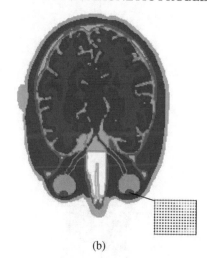

(a)   (b)

FIGURE 4.1: Transverse cut of the discretized human head model for (a) electromagnetic field and (b) thermal effect calculations.

4.2.1.2 The Case of Human Body.

The suitable model for the evaluation, conducted throughout this chapter, lies on the popular data provided by Brooks Air Force Base, San Antonio, TX. Specifically, three distinct resolutions of the discrete model—3, 2, and 1 mm—can be obtained, so leading to the corresponding 3-D grids of $196 \times 114 \times 626$, $293 \times 170 \times 939$, and $586 \times 340 \times 1878$ voxels (Figure 4.2a) [31, 36, 39]. As in the human head, these data are extracted from magnetic resonance images, in every pixel of which a certain type of biological tissue is assigned. Overall, a total of 39 tissue types can be recognized (bile, bladder, blood, blood vessels, body fluid, bone cancellous, bone cortical, marrow, cartilage, cerebro spinal fluid, cornea, lens, retina, dura, vitreous humor, fat, gall bladder bile, glands, heart, small and large intestine, liver, tendons, kidneys, internal and external lung, lymph, mucous membrane, muscles, nails, nerves, white and gray matter, cerebellum, pancreas, skin, spleen, testis, teeth), as depicted in Figure 4.2b.

Having determined the prior materials, the grid for electromagnetic field calculations is constructed through the association of the 12 vortex edges to the constitutive parameters (relative permittivity and electric conductivity) of the tissue located in its center. According to the common practice, at the interface of dissimilar media, the spatial average of their parameters is used. In a similar framework, one can generate the mesh for the thermal analysis, with the only exception that the computational nodes now coincide to the centers of the FDTD cells. This modification is attributed to the different nature of the measured items during the time-domain approach, i.e., electromag-

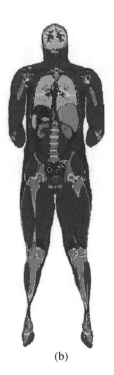

(a)   (b)

FIGURE 4.2: Discrete vortex-based model of the human body. (a) Computational lattice. (b) Distribution of biological tissues along a longitudinal cut.

netic field simulations evaluate vectors, whereas thermal phenomena may be fully described by means of scalar quantities. It should be stressed that in contrast to the high-accuracy stipulations holding for radiating structures (waveguides, antennas, transmission lines, etc.), the analogous constraints for the human head and body are adequately less strict owing to the strong variation of biological quantities. For instance, their exact characteristics and geometry differ considerably among diverse individuals. Even more conspicuous are the discrepancies between the constitutive parameters of biological tissues with regard to frequency (dispersive profile), person, and environmental status, as elaborately stated in the following paragraphs.

From the above, it is evident that the determination of the proper constitutive traits is of crucial importance for the simulation of the human body. As a matter of fact, in scientific literature there exists a multitude of experimental studies for the measurement of relative permittivities and electric conductivities of biological tissues at various frequencies. Based on an detailed survey, the values of the most indicative tissues for the human head—obtained via in vitro measurements—are summarized in Table 4.1 regarding the frequencies of 0.9 and 1.8 GHz (mobile telephony) and 2.44, 5.25, and 5.80 GHz (WLAN antennas).

TABLE 4.1: Relative permittivity and electric conductivity of biological tissues in the human head at the microwave frequency spectrum

| Tissue type | RELATIVE PERMITTIVITY | | | | | ELECTRIC CONDUCTIVITY (S/m) | | | | |
| | FREQUENCY (GHz) | | | | | FREQUENCY (GHz) | | | | |
	0.9	1.8	2.44	5.25	5.80	0.9	1.8	2.44	5.25	5.80
Cornea	55.253	52.768	51.631	47.358	46.533	1.394	1.858	2.288	5.009	5.664
Lens	46.573	45.353	44.636	41.366	40.686	0.855	1.147	1.498	3.807	4.370
Dura	55.271	53.568	52.642	48.626	47.806	1.167	1.602	2.026	4.793	5.466
Nerves	32.351	30.867	30.155	27.68	27.218	0.574	0.843	1.085	2.583	2.942
Vitreous humour	68.902	68.573	68.214	65.5	64.778	1.636	2.033	2.470	5.790	6.674
Cartilage	42.653	40.215	38.792	33.156	32.149	0.782	1.287	1.748	4.336	4.891
Lymph	59.684	58.142	57.215	52.942	52.053	1.039	1.501	1.960	4.984	5.721
Teeth	12.454	11.781	11.387	9.923	9.674	0.143	0.275	0.392	1.022	1.154
Glands	59.684	58.142	57.215	52.942	52.053	1.039	1.501	1.960	4.984	5.721
Cerebellum	49.444	46.114	44.822	40.716	39.98	1.263	1.709	2.095	4.439	4.995
Blood vessels	44.775	43.343	42.544	38.964	38.231	0.696	1.066	2.536	3.781	4.345
Spinal fluid	68.638	67.2	66.258	61.494	60.469	2.413	2.924	3.449	6.975	7.840
Marrow	5.504	5.372	5.298	5.0141	4.963	0.040	0.069	0.095	0.249	0.285
Mucous membrane	46.08	43.85	42.867	39.303	38.624	0.845	1.232	1.586	3.808	4.342
Bone cancellous	20.788	19.343	18.56	15.839	15.394	0.34	0.588	0.802	1.916	2.148
Tendon	45.825	44.251	43.139	37.813	36.749	0.718	1.201	1.677	4.595	5.255

(*Table 4.1 continues*)

	TABLE 4.1: (*continued*)								
	RELATIVE PERMITTIVITY				ELECTRIC CONDUCTIVITY (S/m)				
	FREQUENCY (GHz)				FREQUENCY (GHz)				
White matter	38.886	37.011	36.178	33.186	32.621 0.591	0.915	1.210	3.052	3.494
Gray matter	52.725	50.079	48.928	44.49	44.004 0.942	1.391	1.801	4.370	4.987
Muscle	55.032	53.549	52.742	49.213	48.485 0.943	1.341	1.732	4.323	4.962
Fat	5.462	5.349	5.2811	5.0057	4.954 0.051	0.078	0.104	0.258	0.293
Skin	41.405	38.872	38.018	35.569	35.114 0.867	1.185	1.459	3.259	3.717
Blood	61.36	59.372	58.28	53.512	52.539 1.538	1.066	2.536	5.734	6.506

4.2.2   Frequency-Dependent FDTD Schemes

Because bioelectromagnetic problems exhibit a nontrivial dependency on the selected frequency and the spectrum of our study is rather large (0.9–5.80 GHz), the *frequency-dependent (dispersive) FDTD*—also known as *(FD)²TD*—method is implemented [1–3]. In this technique, all biological tissues are modeled as Debye media with two distinct relaxation times calculated via the nonlinear least squares approach applied to the available measurement data. Essentially, there is a whole class of dispersive schemes for the treatment of such materials. However, in this work the *auxiliary differential equation (ADE) FDTD* algorithm has been selected. Their principal asset lies on the connection of polarization and electric flux density vectors by a complementary time-domain differential equation, updated concurrently with the main time-domain scheme. The development initiates from Ampère's law which, at any grid node in the tissue, becomes

$$\nabla \times \mathbf{H} - \sum_{m=1}^{M} \mathbf{J}_m = \varepsilon_0 \varepsilon_\infty \frac{\partial \mathbf{E}}{\partial t} + \sigma \mathbf{E}, \qquad (4.1)$$

with $\mathbf{J}_m$ as the polarization current assigned to the mth pole of the medium's susceptibility function and ε_∞ the relative permittivity at infinite frequency. In fact, the key aim is to extract an efficient expression for $\mathbf{J}_m$ for its simultaneous update with (4.1). For the sake of brevity and without loss of generality, let us concentrate on the complex polarization current of the mth pole, written as

$$\mathbf{J}_m = \varepsilon_0 \Delta \varepsilon_m \left(\frac{j\omega}{1 + j\omega \tau_m} \right) \mathbf{E} \xrightarrow[\text{Trasnform}]{\text{Fourier}} \mathbf{J}_m + \tau_m \frac{\partial \mathbf{J}_m}{\partial t} = \varepsilon_0 \Delta \varepsilon_m \frac{\partial \mathbf{E}}{\partial t}, \qquad (4.2)$$

where τ_m is the polar relaxation time, $\Delta \varepsilon_m = \varepsilon_{s,m} - \varepsilon_{\infty,m}$ denotes the variation of relative permittivity due to the action of the pole, and $\varepsilon_{s,m}$ is the static relative permittivity. The Fourier-transformed relation in (4.2) is the desired auxiliary differential equation for the time-marching of $\mathbf{J}_m$. Hence, applying the common spatial and temporal central-differencing concepts of the FDTD method, one gets

$$\mathbf{J}_m^{n+1} = A_m \mathbf{J}_m^n + \frac{B_m}{\Delta t} \left(\mathbf{E}^{n+1} - \mathbf{E}^n \right), \quad \text{with} \quad A_m = \frac{1 - 0.5 \tau_m \Delta t}{1 + 0.5 \tau_m \Delta t}, \ B_m = \frac{\varepsilon_0 \Delta \varepsilon_m \Delta t}{\tau_m \left(1 + 0.5 \tau_m \Delta t \right)}, (4.3)$$

As observed, the temporal evolution of $\mathbf{E}^{n+1}$ needs $\mathbf{J}_m^{n+1/2}$ values, resolved from the averaging practice of

$$\mathbf{J}_m^{n+1/2} = 0.5 \left(\mathbf{J}_m^{n+1} + \mathbf{J}_m^n \right) = 0.5 \left[(1 + A_m) \mathbf{J}_m^n + B_m \left(\mathbf{E}^{n+1} - \mathbf{E}^n \right) / \Delta t \right] \qquad (4.4)$$

Therefore, and after some algebra, (4.1) is discretized as

$$\mathbf{E}^{n+1} = \frac{1}{2\varepsilon_0 \varepsilon_\infty + B + \sigma \Delta t} \left\{ (2\varepsilon_0 \varepsilon_\infty + B - \sigma \Delta t) \mathbf{E}^n + 2\Delta t \left[\nabla \times \mathbf{H}^{n+1/2} - \frac{1}{2} \sum_{m=1}^{M} (1 + A_p) \mathbf{J}_m^n \right] \right\},$$

$$(4.5)$$

for $B = \sum_{m=1}^{M} B_m$. Generally, the technique proceeds in a threefold fully explicit manner during a time-step: (a) The components of $\mathbf{E}^{n+1}$ are computed by the already known $\mathbf{E}^n$, $\mathbf{J}_m^n$, $\mathbf{H}^{n+1/2}$ values, (b) the new $\mathbf{J}_m^{n+1}$ components are acquired through (4.3) and the just-updated $\mathbf{E}^{n+1}$, and (c) the future $\mathbf{H}$-field quantities are derived from $\mathbf{H}^{n+1/2}$ and $\mathbf{E}^{n+1}$ vectors in terms of the ordinary FDTD configuration.

Because of its structural complexity, the computational model is further enhanced by means of a subgridding technique, when thin geometries such as wires, helices, and patches are to be discretized. In particular, the fine mesh is placed within the coarse one (Figure 4.3), such that its boundaries are as close as possible but not coinciding to those of the dual coarse grid. Apart from the standard FDTD update schemes in both lattices, all components of the fine counterpart, which are in the transition region and cannot be updated, are expressed via an accurate spline approximation, using three components from the coarse mesh in each direction [19]. On the other hand, coarse-grid components, overlapped by fine-grid ones, are simply interpolated, whereas for any missing temporal component, extrapolation techniques are incorporated. The technique, combined also with an eight-cell PML, is stable and the memory requirements are not significantly increased.

The aforesaid computation procedure leads to the values of antenna efficiency and SAR, given by

$$\alpha = \frac{1}{P_{\text{in}}} \left(P_{\text{in}} - \int_{\text{head}} \sigma |E|^2 dv \right) \quad \text{and} \quad \text{SAR} = \frac{\sigma |E|^2}{2\rho} \tag{4.6}$$

for P_{in}, the input power, and ρ, the density of the material, under examination. Finally, SAR values are postprocessed to yield the average quantities in alignment with the international safety standards.

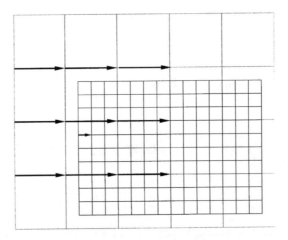

FIGURE 4.3: Location of the subgrid with regard to the coarse mesh. Any component in the transition region (short arrow) is updated via three components from the coarse lattice in each direction (long arrows).

4.2.3 Compliance With International Safety Standards

The ongoing public demand for the establishment of general safety standards emerged from the vast amount of questions concerning the hazard levels of human exposure to electromagnetic fields radiated by various sources, such as high-voltage transmission lines, cellular phones, microwave ovens, radio, and television antennas as well as WLAN devices. Consequently, a set of international rules, recommendations, and thresholds have been determined, the most important of which are:

- The *IEEE C95.1 (1999)*. This standard—undergoing a thorough reestablishment—adopts the SAR indicator in the zone of 0.1 MHz to 6 GHz, although for larger frequencies it opts for the power density [48].
- The *International Commission on Non-Ionizing Protection (ICNIRP) (1998)*. This standard suggests certain limitations with respect to SAR in the spectrum of 10 kHz to 10 GHz and relative to power density for larger frequencies [49].
- The *Council Recommendation of the European Community 1999/519/EC (1999)* referring to general public safety. It is completely harmonized with the ICNIRP standard [50].
- The *Comité Européen de Normalisation Electrotechnique Standards (2001, 2002)*. They have been published in the Official Journal of the European Communities and are totally coordinated with the 1995/5/EC recommendation [51, 52].
- The *Recommendations of the Canadian Safety Code 6 Standard (1999)* that are similar to the IEEE C95.1 Standard [53].

Despite the sometimes distinguishable discrepancies among the above standards, all international associations conduct a serious effort to smoothen such differences by introducing extra amendments and modifications.

4.3 BIOLOGICAL EFFECTS OF CELLULAR PHONE RADIATION

One of the most critical human exposure cases to microwave power is accredited to the usage of cellular phones, a fact that is substantiated by the intense research activities studying the resultant absorption levels. Because the radiating device is located very close in the human head, there is always the serious possibility for some local SAR values to be rather high, even when the mean of the quantity remains tolerably low in the entire head. Furthermore, due to the relatively short duration of cellular telephone calls, the particular exposure can be deemed extremely fast, thus leaving the transient phenomenon, responsible for the temperature rise, incomplete. For these reasons, in the next paragraphs, some indicative, yet typical, situations are analyzed.

4.3.1 Discrete Models of Various Cellular Phone Types

The first generic model of a cellular phone antenna is the quarter-wavelength dipole, mounted on a conducting box, as shown in Figure 4.4a. This assumption is justified by the presence of the PCB and battery, but it is appropriately altered for other phone types to yield more rigorous discretizations. The second model (Figure 4.4b) is based on the more common helical antenna. Such devices, for small values of their diameter and length, result in nearly isotropic radiation on the horizontal plane. The third and fourth models correspond to modern types of phones with patch antennas. This may enhance the phone's appearance, but entails very careful design to produce the desirable radiation pattern. The side-mounted PIFA (Figure 4.4c) and another patch antenna model (Figure 4.4d) are amid the most common to be implemented [19, 24].

Analytically, the quarter-wavelength monopole is mounted at the top of the cellular phone so that it does not touch the human head. In the simulations, the frequencies of 900 and 1800 MHz are used, whereas the total radiation power is set to the rather high values of 600 and 250 mW, respectively. Similar considerations hold for the helical antenna as well. Proceeding to contemporary arrangements, the internally located patch antenna enabled the fabrication of more compact and attractive structures [22, 32, 42]. Except for the reduction of their overall volume, such devices are user-friendly and exhibit larger versatility and shock durability. In addition, they are proven much safer—compared to the classical cellular phones—in the absorption of electromagnetic power from the human head. For the discretization of the latter phone, special attention is drawn on its PCB

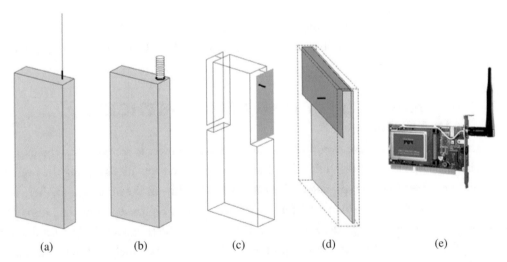

(a)   (b)   (c)   (d)   (e)

FIGURE 4.4: Cellular phone and WLAN antennas. (a) Quarter-wavelength monopole, (b) helix antenna, (c) side-mounted PIFA, (d) patch antenna, and (e) WLAN antenna. Feeds are shown with a thick line.

card, the short circuit between the antenna and card, as well as on the dielectric substrate ($\varepsilon_r = 2$) of the whole radiator. Although the realistic model of the cellular phone depends on more parameters, such as the feed position, the volume of cover, or the interaction of the battery with the antenna, the prior setup constitutes a very reliable approximation. Figure 4.5a and 4.5b provides a set of typical dimensions for a 900- and an 1800-MHz cellular phone, respectively. The validity of these models may be promptly deduced through Figure 4.6, which illustrates their reflection coefficients in the case of free-space operation.

The last type of cellular phone is equipped with a couple of side-mounted PIFA structures that offer small dimensions, large bandwidth, satisfactory gain, and multichannel function. In its simple rendition, a PIFA consists of a microstrip patch with an electrical size of $\lambda/4$, short-circuited in its one end. In this manner, a prefixed resonance frequency is accomplished with half the actual size of the antenna. A normal computational model is shown in Figure 4.5c, where two PIFAs are used for the attainment of a multiple spatial reception and the suppression of multipath parasites. The formulation, so developed, allows the use of one or both PIFAs as radiators; however, their concurrent operation has been proven detrimental for the efficiency of the phone. Lastly, the location of the antennas' excitation point is chosen in order to achieve the correct matching between the radiation and the internal impedance of the feed circuit.

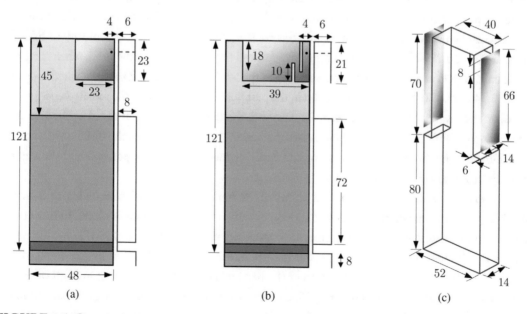

(a)   (b)   (c)

FIGURE 4.5: Geometric representation of three cellular phones with (a) a patch antenna at 900 MHz, (b) a patch antenna at 1800 MHz, and (c) a PIFA at 1800 MHz. All dimensions are in millimeter (mm) and the radius of the two holes (black dots) is 1 mm.

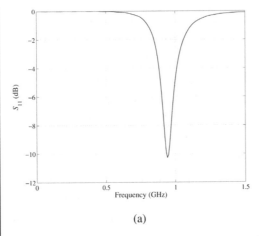
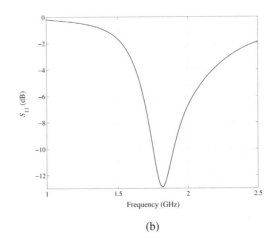

(a)   (b)

FIGURE 4.6: Numerical calculation of the reflection coefficient for the patch-antenna cellular phone at (a) 900 MHz and (b) 1800 MHz.

4.3.2 Comparative Study—Numerical Results

In this paragraph, the absorption of electromagnetic radiation from various types of cellular phone antennas, described above, is comprehensively investigated with an emphasis on human head. The primary quantity computed in all FDTD simulations is SAR, which is considered the most significant and characteristic indicator for such problems. In this context, Figures 4.7 and 4.8 present various SAR images, at 900 and 1800 MHz, in the coronal, frontal, and sagittal sections of the head, when the phone antenna is located at a distance of 3 and 6 mm from the left ear, respectively. Different "tilted" positions (i.e., 30°, 45°, and 60°) are examined to obtain the best possible estimation of the radiated energy influences on the diverse tissues of the head. As observed, the parts of the head situated in the vicinity of the antenna exhibit the larger absorption rates that are then degenerate with the distance in a smooth way. Furthermore, from the outcomes one can easily deduce that both the position and the antenna type drastically affect the SAR strength.

Next, two additional illumination points of the human head are selected owing to their vulnerability in electromagnetic radiation. More specifically, the cellular phone is placed in front of the left eye and the right ear. The simulations are conducted at both 900 and 1800 MHz to sufficiently resolve the differences in the absorption mechanism of the energy. To this aim, Figure 4.9 illustrates several SAR images in the coronal section of the head when a $\lambda/2$ antenna is set in front of the left eye at the consecutive distances of $d = 5, 15, 25$ mm. Note that all SAR values are normalized to the maximum (worst) $d = 5$ mm case. From the results, it becomes apparent that as operation frequency increases, the penetration of propagating waves is considerably limited. Figure 4.9 also reveals the nontrivial possibility of a rather serious concentration of the absorbed power in the area of the eye, an issue that justifies our concern for the particular organ. Actually, the geometry of the eye favors

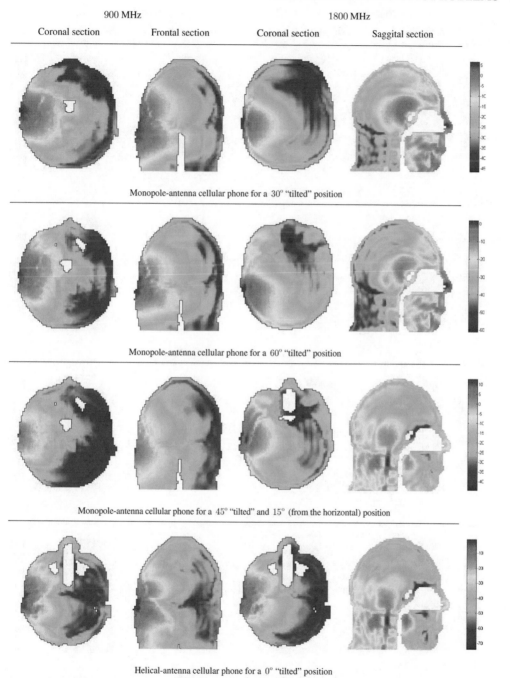

FIGURE 4.7: 900 and 1800 MHz SAR images in decibels (W/kg) in the coronal, frontal, and sagittal sections of the human head for different cellular phones and "tilted" positions at a 3-mm distance from the left ear.

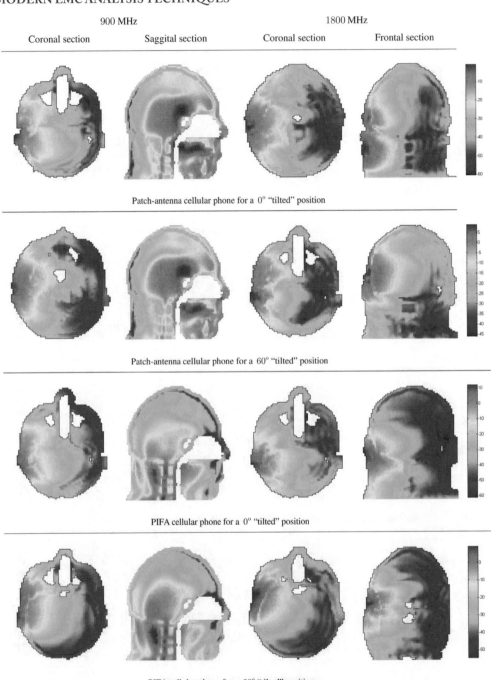

FIGURE 4.8: 900 and 1800 MHz SAR images in decibels (W/kg) in the coronal, sagittal, and frontal sections of the human head for different cellular phones and "tilted" positions at a 6-mm distance from the left ear.

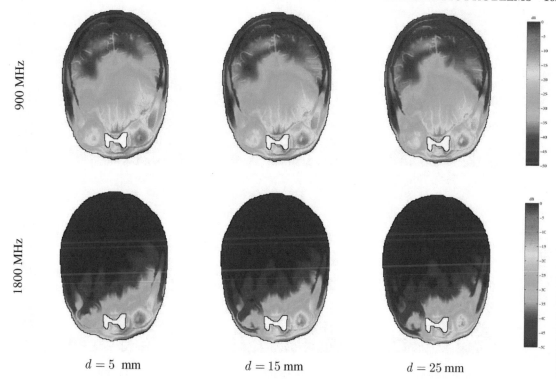

$d = 5$ mm $d = 15$ mm $d = 25$ mm

FIGURE 4.9: 900 and 1800 MHz SAR images in the coronal section of the human head for a $\lambda/2$ dipole-antenna cellular phone located in front of the left eye. All values are normalized with regard to the maximum (worst) case, $d = 5$ mm.

the appearance of hot spots, especially at 1800 MHz. In a similar fashion, Figure 4.10 displays the normalized SAR distribution in the frontal section of the head, when the same $\lambda/2$ antenna is placed in front of the right ear at $d = 2, 12, 22$ mm. Again, deeper penetration depths are attained at 900 MHz, while the ear is subject to the largest portion of the absorbed power. It should be stressed that the FDTD method has been proven very efficient and accurate in the solution of these demanding problems, hence justifying their important role in bioelectromagnetic applications.

Regarding electric field intensity, Figure 4.11 gives an assortment of indicative 900 and 1800 MHz snapshots for a monopole, a helical, and a patch cellular phone antenna in the coronal and frontal sections of the head. The conclusions derived from SAR distributions are valid here as well without any noticeable variations.

Moving to a double PIFA device, the analysis explores its behavior in the case of 3-mm distance from the head for a 0° and a 30° "tilted" position at 900 and 1800 MHz. The specific distance is deemed among the worst ones for human exposure, because the thickness of the phone's dielectric cover is usually larger than the selected value of 3 mm. The normalized SAR distribution for these

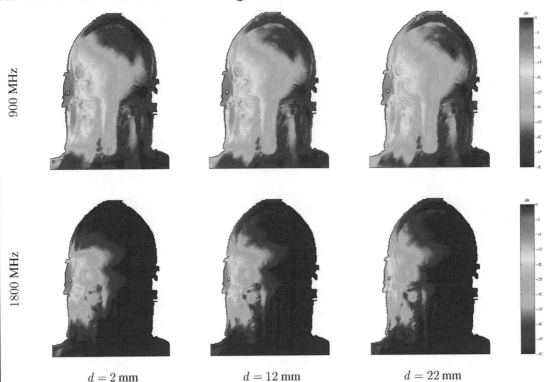

FIGURE 4.10: 900 and 1800 MHz SAR images in the frontal section of the human head for a $\lambda/2$ dipole-antenna cellular phone located in front of the right ear. All values are normalized with regard to the maximum (worst) case, $d = 2$ mm.

cases is shown in Figure 4.12, from where the reduction of the penetration depth with the increase of operation frequency may be realized. Note that for the 900-MHz arrangement, the output power of the structure is equal to 250 mW, whereas for the 1800 MHz setup is 125 mW. These realistic values are based on the fact that according to the Global System for Mobile Communication (GSM) protocol, such phones function with a maximum radiated power of 2 and 1 W, respectively [32]. As a typical user, can occupy only one of the eight available system channels, the equivalent output power is $2 \div 8 = 0.25$ and $1 \div 8 = 0.125$ W.

To this extent, Table 4.2 summarizes the maximum SAR values on the skin, when the cellular phone is 3 mm away from the head. Moreover, for comparison, the respective SAR_{1g} and SAR_{10g} indicators (SAR per 1 and 10 g of tissue) are included. For both, the regions involved in the extraction of their mean values are cubes. In the case of SAR_{10g}, in particular, this region is allowed to contain up to the 20% of the total air volume. Delving into the outcomes of the table, one can discern a notable discrepancy in the SAR at the two frequencies. In contrast, the smallest wavelengths

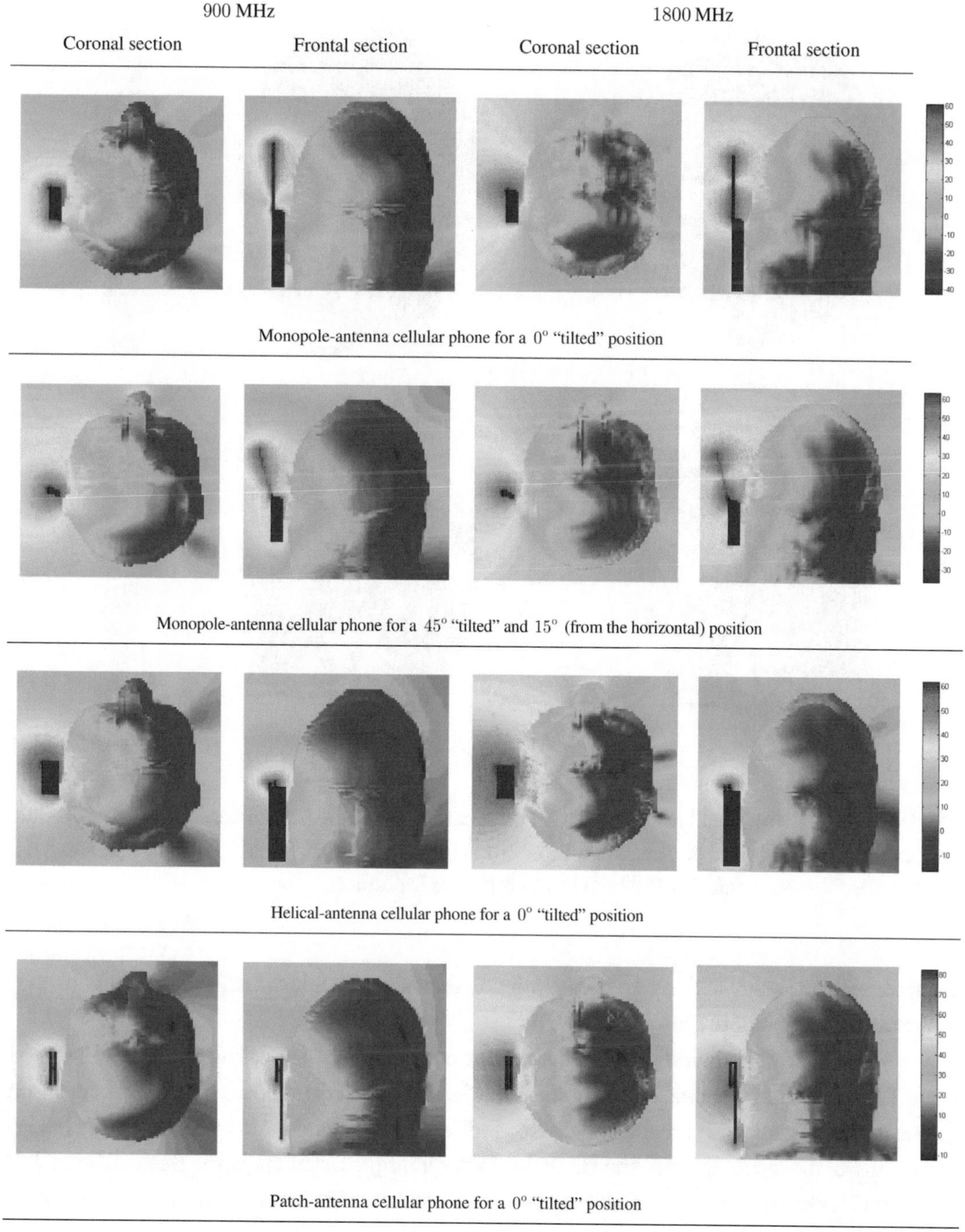

FIGURE 4.11: 900 and 1800 MHz electric field snapshots in decibels (V/m) in the coronal and frontal section of the human head for different cellular phones and "tilted" positions at a 3-mm distance from the left ear.

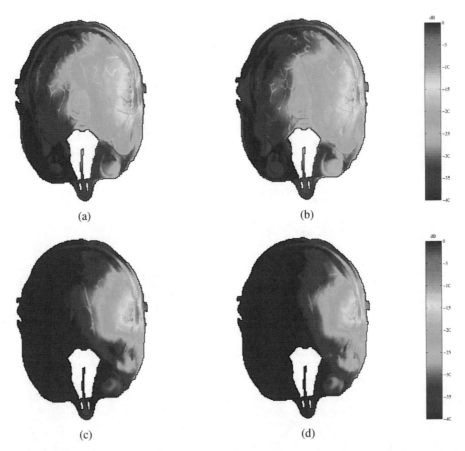

FIGURE 4.12: Normalized SAR images in the coronal section of the human head from a PIFA cellular phone at (a) 900 MHz and 0° "tilted" position, (b) 900 MHz and 30° "tilted" position, (c) 1800 MHz and 0° "tilted" position, and (d) 1800 MHz and 30° "tilted" position.

are accompanied by smaller penetration depths, as shown at the SAR distributions in Figure 4.13 in the frontal section of the head.

Subsequently, the investigation moves to the comparison of the calculated SAR values with those dictated by the international safety standards in an effort to comprehend the levels of exposure consequences. Figure 4.14a presents the SAR_{1g} indicator with regard to distance d from the left eye and compares the outcomes with the IEEE 1.6 W/kg limit [48]. Although, the frequencies of 2.44 and 5.80 GHz are studied in the following section, their values are included in the plots for convenience and generality. It is observed that when the cellular phone is located very close to the head, there is a large possibility for the limit to be surpassed, despite the small values of the total radi-

TABLE 4.2: Maximum SAR_{1g} and SAR_{10g} values for a PIFA cellular phone

FREQUENCY (MHz)	TILT (DEG)	POWER (mW)	RADIATED POWER (%)	SAR_{1g}, SKIN	SAR_{1g}, BRAIN	SAR_{10g}
900	0	250	32.55	1.293	0.361	0.752
900	30	250	30.66	1.140	0.262	0.775
1800	0	125	46.64	1.571	0.334	0.757
1800	30	125	45.58	1.627	0.308	0.778

ated power. In the example of Figure 4.14a, this marginal distance is 7–10.5 mm. Also, it becomes evident that SAR_{1g} dwindles with the increase of operation frequency. On the other hand, Figure 4.14b depicts the variation of SAR_{10g} indicator versus distance d from the left eye and compares the results with the ICNIRP 2 W/kg limit [49]. Here, the maximum value (1.307 at 2.44 GHz) is much lower than the prescribed threshold.

Analogous simulations are conducted for the case of a cellular phone in front of the right ear. Likewise, Figure 4.14c provides SAR_{1g} and examines possible violations of the IEEE limit [48]. Under these circumstances, for a safe usage of the $\lambda/2$ dipole-antenna device, a distance of 13.3 mm or greater is required. Nevertheless, it must be clarified that the aforesaid maximum values are basically

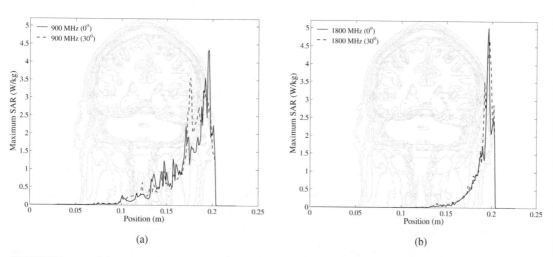

(a)   (b)

FIGURE 4.13: Maximum SAR values in the frontal section of the human head for a patch-antenna cellular phone at (a) 900 MHz, (b) 1800 MHz and different "tilted" positions.

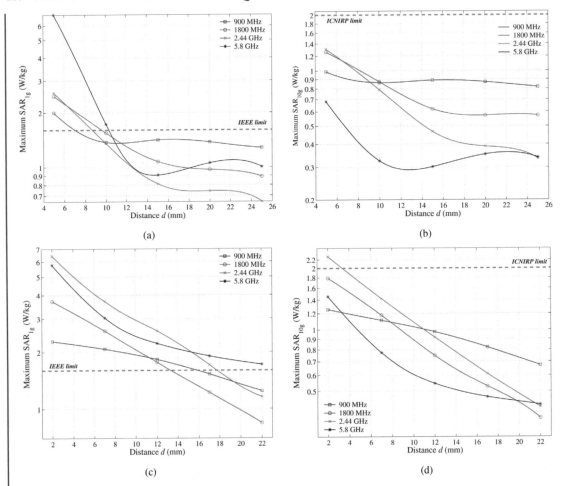

FIGURE 4.14: Variation of SAR mean values versus the distance of a $\lambda/2$ dipole-antenna cellular phone from the left eye: (a) SAR_{1g} case (IEEE limit), (b) SAR_{10g} case (ICNIRP limit); and from the right ear: (c) SAR_{1g} case (IEEE limit), (d) SAR_{10g} case (ICNIRP limit). For comparison, the values of frequencies 2.44 and 5.8 GHz are, also, provided.

experienced at the lobe of the ear, which is likely to be considered a marginal part of the human body in a future reformulation of the IEEE safety standard (i.e., higher allowable local SAR compared to the rest of the body). Furthermore, SAR_{1g} is much smaller in the vicinity of the brain than on the skin. To complete our verifications, Figure 4.14d shows the SAR_{10g} indicator with respect to the ICNIRP limit [49]. Again, no serious violations are observed with the only exception the operation at 2.44 GHz for values lower than 2.85 mm. In particular, the lowest values of SAR_{10g} are obtained at 5.80 GHz, a fact that must be probably accredited to the very small penetration depth in the human body.

Finally, a detailed study on the various cellular phone antennas is performed at the frequencies of 900 and 1800 MHz. Two positions are examined: the vertical and the very typical 30° "tilted" one. Table 4.3 gives the results for diverse quantities, such as SAR, SAR_{1g}, and SAR_{10g} indicators

TABLE 4.3. Maximum SAR_{1g} and SAR_{10g} and efficiency values for different cellular phone antennas

ANTENNA TYPE	FREQUENCY (MHz)	TILT (DEG)	MAXIMUM SAR	MAX AVERAGE SAR_{1g}	MAXIMUM AVERAGE SAR_{10g}	ANTENNA EFFICIENCY (%)
Monopole	900	0	7.993	3.181	2.061	25.29
		30	4.596	2.925	2.072	30.40
	1800	0	2.171	1.473	0.946	64.98
		30	2.013	1.141	0.689	64.80
Helix	900	0	7.255	4.011	2.531	54.36
		30	5.771	3.477	2.385	62.21
	1800	0	3.838	8.550	5.402	46.07
		30	3.347	6.621	4.097	53.21
PIFA	900	0	13.852	5.305	3.299	20.41
		30	6.078	3.902	2.430	29.57
	1800	0	16.032	8.769	4.565	23.78
		30	8.372	5.397	3.302	33.43
Patch	900	0	7.819	4.764	3.178	28.50
		30	8.622	5.022	3.105	31.37
	1800	0	1.152	0.855	0.591	66.33
		30	0.808	0.626	0.447	71.82

and the antenna efficiency. Based on inspection, several fairly important differences in the maximum values of the prior quantities are discerned, depending on the antenna kind. Specifically, the "tilted" position yields milder outcomes in all cases, because of the relatively remote location of the device from the human head. Contrarily, antenna efficiency is significantly augmented at 1800 MHz, an anticipated deduction mainly attributed to the reduced penetration depth inside the tissues. However, for a sufficiently more (if not absolutely) comprehensive judgment on this topic, the thermal analysis, described in the next paragraph, should be performed.

4.3.3   Examination of Thermal Effects

As can be deduced from the above results, electromagnetic analysis gives only a coarse estimation of how the power is absorbed by the tissues and thus cannot account for the exact description of thermal phenomena. Actually, experimental research on animals has shown that the hazardous influence of electromagnetic radiation is principally accredited to the temperature rise in the tissues [16, 21, 30, 36]. To this extent, the heat of a tumor comprising biological materials is affected by several factors, such as the distribution of the absorbed power, the values of material thermal parameters, the existence of discontinuities, and the rate of temperature exchange between the human body and the surrounding environment [18, 23, 35, 41]. Consequently, a more elaborate examination of thermal effects should be pursued to obtain a precise picture of power absorption.

4.3.3.1 Theoretical Formulation. Coherent to these stipulations is the analysis based on the bio-heat equation

$$K\nabla^2 T + A_0 + \rho \mathrm{SAR} - B(T - T_b) = C\rho\frac{\partial T}{\partial t}, \tag{4.7}$$

which equals the heat accumulated (or lost) at a specific spot per unit volume and time to the temporal variation of temperature, multiplied by the thermal capacity in a 1 m³ volume of the tissue under study. In (4.7), thermal capacity is given via the product of specific heat C (J/(kg °C)) and density ρ (kg/m³) of the tissue. In particular, the left-hand side of (4.7) includes all possible ways of heat production, transfer, or consumption at a certain point. Moreover, K (J/(s m °C)) is the thermal conductivity, A_0 (J/(s m³)) the per unit volume power generated by metabolism, and B (J/(s °C m³)) represents the mechanism of thermal distribution T related to blood flow and the deviation of temperature from that of blood T_b. Table 4.4 and Figure 4.15, respectively, give the most characteristic values and spatial distribution of these thermal parameters for the tissues of Table 4.1 along with their corresponding density [3–11].

TABLE 4.4: Thermal Parameter and Density Values of Various Tissue Types

TISSUE TYPE	C (J/(kg °C))	K (J/(s m °C))	A_0 (J/(s m³))	B (J/(s °C m³))	ρ (kg/s³)
Cornea	4178	0.58	0	0	1076
Lens	3000	0.40	0	0	1053
Dura	4200	0.58	0	0	1026
Nerves	3500	0.46	7100	40,000	1038
Vitreous humor	3997	0.59	0	0	1008.9
Cartilage	3500	0.47	1600	9000	1097
Lymph	3840	0.53	0	0	1040
Teeth	3600	0.53	7000	25,000	1050
Glands	3700	0.57	7100	40,000	1038
Cerebellum	4200	0.62	0	0	1007.2
Blood vessels	2700	0.22	5700	32,000	1040
Spinal fluid	3300	0.43	1600	9000	1040
Marrow	1300	0.40	590	3300	1920
Mucous membrane	3500	0.50	690	2700	1220
Bone cancellous	1300	0.40	610	3400	1990
Tendon	3600	0.50	10,000	40,000	1038
Bone cortical	3700	0.57	7100	40,000	1038
White matter	3600	0.50	480	2700	1046.85
Gray matter	2500	0.25	300	1700	1022
Muscle	3500	0.42	1620	9100	1125
Fat	3300	0.43	1600	9000	1040

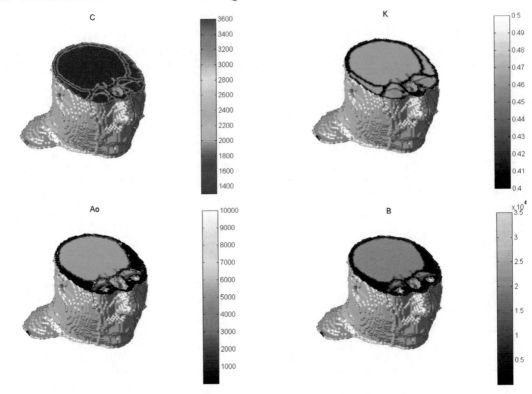

FIGURE 4.15: Distribution of the four thermal parameters involved in the bioheat equation in a coronal section of the human head.

Of critical significance is the boundary condition—derived from the continuity of heat flow along the normal outward direction—to be enforced at the exterior surface of the skin and internal cavity, expressed as

$$K(\nabla T \cdot \hat{\mathbf{n}})_S = h(T - T_a), \tag{4.8}$$

where S denotes the association of surface points, h is the convective heat transfer coefficient ($J/(m^2\ s\ °C)$), T_a is the ambient temperature, and $\hat{\mathbf{n}}$ is a unit vector perpendicular to the outer surface. Note that for the majority of the results, throughout this chapter, h has been set to the popular value of 10.5 $J/(m^2\ s\ °C)$.

Nevertheless, in the present analysis a slightly *altered* approach is introduced in order to exclusively focus on the time-domain calculation of thermal rise, due to power absorption, and omit other factors that contribute to the overall thermal equilibrium. Thus, considering the resultant temperature distribution T as the sum of thermal equilibrium distribution T_0 (without the cellular phone) and temperature rise $\Delta T = T - T_0$, the corresponding expressions will be similar to (4.7) and (4.8), with the sole exception that the A_0 term will be absent. According to these conventions, the bioheat equation and the convective boundary condition become

$$K\nabla^2\Delta T + \rho\mathrm{SAR} - B\Delta T = C\rho\frac{\partial \Delta T}{\partial t}, \qquad (4.9)$$

$$K(\nabla\Delta T \cdot \hat{\mathbf{n}})_S = h\Delta T \qquad (4.10)$$

In this manner, the power due to metabolism, and the blood and ambient temperatures, do not affect the temperature rise in the computational model. Equation (4.9) can be discretized by means of an Euler conditionally stable time-domain scheme using central differences for the Laplacian and forward differences for the temporal derivative, on the condition that the maximum allowable time-step is not prohibitive for an efficient implementation. Therefore, the discrete form of (4.9) for a cubic mesh ($\Delta x = \Delta y = \Delta z = \Delta$) is

$$
\begin{aligned}
\Delta T|_{i,j,k}^{n+1} = &\left(1 - \frac{B_{i,j,k}\Delta t}{C_{i,j,k}\rho_{i,j,k}} - \frac{6K_{i,j,k}\Delta t}{C_{i,j,k}\rho_{i,j,k}\Delta^2}\right)\Delta T|_{i,j,k}^{n+1} \\
&+ \frac{K_{i,j,k}\Delta t}{C_{i,j,k}\rho_{i,j,k}\Delta^2}\left(\begin{array}{l}\Delta T|_{i+1,j,k}^{n} + \Delta T|_{i,j+1,k}^{n} + \Delta T|_{i,j,k+1}^{n} \\ + \Delta T|_{i-1,j,k}^{n} + \Delta T|_{i,j-1,k}^{n} + \Delta T|_{i,j,k-1}^{n} - 6\Delta T|_{i,j,k}^{n}\end{array}\right) \\
&+ \frac{\Delta t}{C_{i,j,k}}\mathrm{SAR}_{i,j,k}
\end{aligned} \qquad (4.11)
$$

To validate the stability of (4.13), the well-known von Neumann analysis after some mathematical manipulations leads to the criterion of

$$\Delta t \le \frac{2C\rho\Delta^2}{12K + B\Delta^2} \simeq \frac{C\rho\Delta^2}{6K}, \qquad (4.12)$$

which, for a cell size of 3 mm, yields a maximum allowable temporal increment of 5.83 s, i.e., less than 500 iterations, even for a 30-min exposure duration.

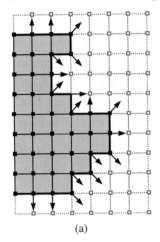 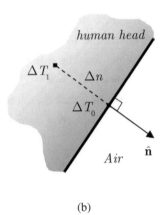

(a)   (b)

FIGURE 4.16: Quasi-conformal representation of the (a) exterior human body and internal cavity surface and (b) discretization scheme for the convective heat transfer boundary condition.

Conversely, for the convective boundary condition a more systematic discretization strategy must be adopted. Therefore, unlike existing simplified practices, in this work a quasi-conformal representation of the exterior body surface is launched, as shown in Figure 4.16a. According to this technique, all nodes belonging to the exterior surface should be clearly identified. Such a prerequisite can be fulfilled by checking if the node under study is directly connected to (at least) one node of the free space [19]. The exact number of the latter is indeed very important, because it determines parameter Δn. In particular, if a surface node is connected to one, two, or three free-space nodes, then $\Delta n = \Delta$, $\Delta n = \sqrt{2}\Delta$, or $\Delta n = \sqrt{3}\Delta$, respectively. Also, the normal vector in Figure 4.16b follows the shape of the head and is oriented toward either x-, y-, or z-axis, or some of the possible xy, yz, zx, xyz diagonal directions. In general, the discretized form of (4.10) will advance temperature terms at any surface node ΔT_0 through the analogous terms at the neighboring node ΔT_1. Thus, one can write

$$\Delta T_0 \big|^{n+1} = \frac{K}{K + h\Delta n} \Delta T_1 \big|^{n+1} \tag{4.13}$$

Whereas the interior diffusion scheme is used to update all values in the tissues, including those at nodes adjacent to the surface, the discrete form (4.10) of the convective boundary condition updates the surface values and the algorithm proceeds with the next time-step. The entire method incorporates a large amount of degrees of freedom and requires only a few minutes to reach the steady-state thermal rise distribution.

FIGURE 4.17: 900 and 1800 MHz thermal rise distributions in decibels (1°C) in the coronal, frontal, and sagittal sections of the human head for different cellular phones and "tilted" positions at a 3-mm distance from the left ear.

900 MHz   1800 MHz

Coronal section   Frontal section   Coronal section   Saggital section

Monopole-antenna cellular phone for a 45° "tilted" and 15° (from the horizontal) position

Helical-antenna cellular phone for a 0° "tilted" position

Patch-antenna cellular phone for a 60° "tilted" position

PIFA cellular phone for a 60° "tilted" position

4.3.3.2 Thermal Simulations. In the following, a set of informative numerical outcomes based on the computational model of the previous paragraph, equipped with the modified convective boundary condition, is presented. Consequently, Figure 4.17 illustrates the thermal rise distribution at 900 and 1800 MHz in the coronal, frontal, and sagittal sections of the human head, as produced by several cellular phones and "tilted" positions in front of the left ear. The metric notation of decibels with regard to 1°C has been chosen for the best possible depiction of the calculated terms, because temperature rise distribution is linearly related to SAR.

Notice the smoothness of the results that reveals the capability of the selected method to successfully treat this demanding kind of EMC problems. Toward this direction, the temperature variations encountered at coronal planes of maximum temperature rise T_{max}, produced by a $\lambda/2$ cellular phone antenna at a distance of $d = 5, 15, 25$ mm in front of the left eye and normalized to the worst ($d = 5$ mm case), are shown in Figure 4.18. Similarly, Figure 4.19 gives the corresponding temperature variations in frontal sections of the head, when the same radiating device is placed at $d = 2, 12, 22$ mm in front of the right ear. This time, normalization is conducted with reference to the worst case

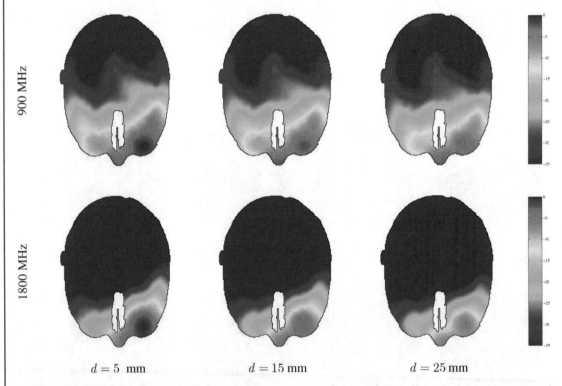

$d = 5$ mm $d = 15$ mm $d = 25$ mm

FIGURE 4.18: 900 and 1800 MHz temperature variations $(10 \log_{10}(T/T_{max}))$ in the coronal section of the human head for a $\lambda/2$ dipole-antenna cellular phone located in front of the left eye. All values are normalized with regard to the maximum case, $d = 5$ mm

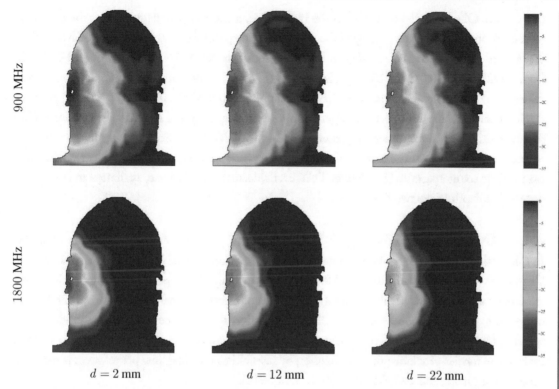

FIGURE 4.19: 900 and 1800 MHz temperature variations $(10 \log_{10}(T/T_{max}))$ in the frontal section of the human head for a $\lambda/2$ dipole-antenna cellular phone located in front of the right ear. All values are normalized with regard to the maximum case, $d = 2$ mm.

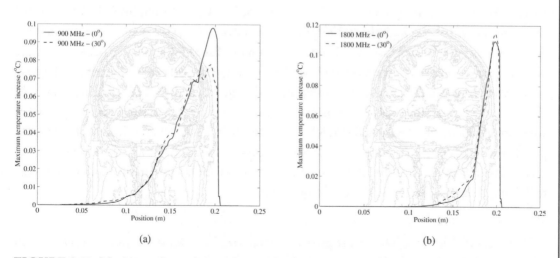

FIGURE 4.20: Maximum thermal rise values in the frontal section of the human head for a PIFA cellular phone at (a) 900 MHz, (b) 1800 MHz and two "tilted" positions.

($d = 2$ mm). Observing the outcomes of the two figures, it can be easily deduced that the maximum rise occurs on the skin in the vicinity of the eye (limits: from 0.135°C to 0.394°C for 900 MHz; from 0.176°C to 0.426°C for 1800 MHz) and the ear (limits: from 0.147°C to 0.418°C for 900 MHz; from 0.192°C to 0.484°C for 1800 MHz). Finally, Figure 4.20 presents the maximum thermal rise values in the frontal head section, for a PIFA cellular phone located at a 0° and a 30° "tilted" position. Again, it becomes apparent that the larger the operation frequency, the closer to the exterior surface the absorbed power concentrates. For example, for both 900 MHz positions, the maximum values are reduced up to the 50% of the total maximum rise after 3.7 and 4.63 cm, compared to 1.63 and 1.61 cm for their counterparts at 1800 MHz. This drastic decrease is, of course, attributed to the smaller penetration depth of the two frequencies.

4.4 INVESTIGATION OF WIRELESS LOCAL AREA NETWORKS

In this section, the influence of electromagnetic radiation from WLAN arrangements, described in Chapter 3, is examined both from a SAR and a thermal analysis viewpoint. Although such devices are usually situated in relatively remote (in the sense of electric terms) distances from the human body—unlike cellular phones—and therefore their action should be considered in a far-field perspective, for the sake of comparison near-field simulations are also included [36–47]. It is stressed that for the former problems, the illumination by far-field electromagnetic plane waves is modeled through the total/scattered field excitation technique [2].

4.4.1 SAR Calculations

4.4.1.1 Near-field Sources. Starting from near-field investigations and assuming a $\lambda/2$ dipole antenna mounted on the WLAN system, Figure 4.21 displays the 2.44 and 5.80 GHz SAR distribution in the coronal section of the head, when the device is placed at a distance of $d = 5, 15, 25$ mm in front of the left eye. As in the analogous cellular phone cases, the outcomes are normalized to the maximum values at $d = 5$ mm. Their inspection unveils the expected reduction of electromagnetic field penetration with the increase in operation frequency. Akin deductions are drawn for the right ear at $d = 2, 12, 22$ mm in Figure 4.22. Furthermore, antenna efficiency (%) as a function of distance and frequency is explored in Figure 4.23, where for generality the corresponding values at 900 and 1800 MHz are provided. As observed, the percentage of the energy absorbed by the tissues is minimized when the involved wavelength is decreased. Essentially, the efficiency of the dipole antenna at 5.80 GHz is practically constant and equal to 80% approximately, except for the $d = 5$ mm distance.

4.4.1.2 Far-field Sources. The next stage of our computational study deals with the time-domain modeling of far-field sources. Because in reality the human body can be illuminated by electromag-

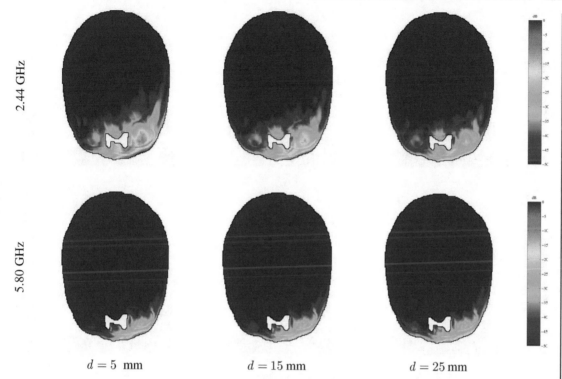

2.44 GHz

5.80 GHz

$d = 5$ mm $d = 15$ mm $d = 25$ mm

FIGURE 4.21: 2.44 and 5.80 GHz SAR distributions in the coronal section of the human head for a $\lambda/2$ dipole-antenna cellular phone located in front of the left eye. All values are normalized with regard to the maximum (worst) case, $d = 5$ mm.

netic waves propagating from different directions—due to multiple scattering interactions with furniture, walls, ceiling, floor, or other individuals—it has been deemed instructive to conduct the simulations for diverse incident angles. For this goal, far-field illumination is realized through plane waves traveling on the horizontal ($\theta = 90°$) plane along directions varying from $-90°$ to $90°$. In this context, Figure 4.24 depicts a set of 2.44 and 5.80 GHz SAR images (normalized to the maximum value of every case) in the coronal section of the head for an exposure to waves on the $\theta = 90°$ plane with diverse incident angles φ. In addition, Figure 4.25 gives the normalized SAR distribution on the surface of regions related to the bones, eyes, and the brain of the head for $\varphi = 90°$. Nonetheless, a more elaborate comparison of the maximum SAR, SAR_{1g}, and SAR_{10g} values, when the power density of the incident radiation is equal to the IEEE (for SAR_{1g}) [48] and ICNIRP (for SAR_{10g}) [49] safety limits in a nonmonitored environment and a general public exposure, respectively, is presented in Table 4.5. For convenience, its results—concerning the overall, the brain, and the eyes case—are also plotted in Figure 4.26 at the frequencies of 2.44 and 5.80 GHz. Focusing on the former frequency, the

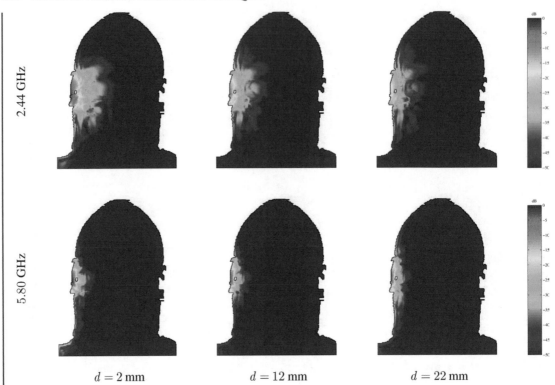

$d = 2$ mm $d = 12$ mm $d = 22$ mm

FIGURE 4.22: 2.44 and 5.80 GHz SAR distributions in the frontal section of the human head for a $\lambda/2$ dipole-antenna cellular phone located in front of the right ear. All values are normalized with regard to the maximum (worst) case, $d = 2$ mm.

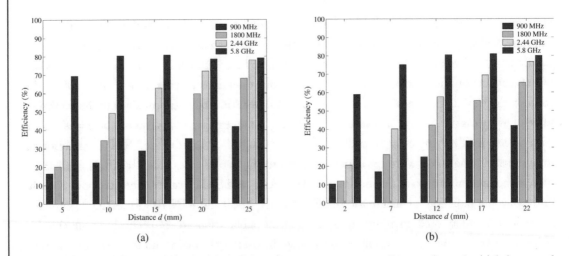

(a)   (b)

FIGURE 4.23: Efficiency (%) of a $\lambda/2$ cellular-phone antenna versus distance from the (a) left eye and (b) right ear. For comparison, the values at 900 and 1800 MHz are also provided.

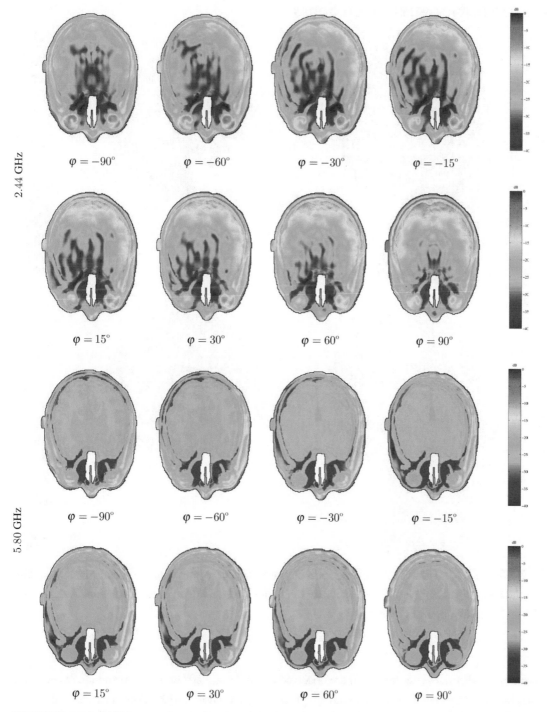

FIGURE 4.24: SAR images (normalized to the maximum value of each case) in the coronal section for human exposure to a 2.44- and a 5.80-GHz electromagnetic plane wave propagating on the $\theta = 90°$ plane with different values of incidence angle φ.

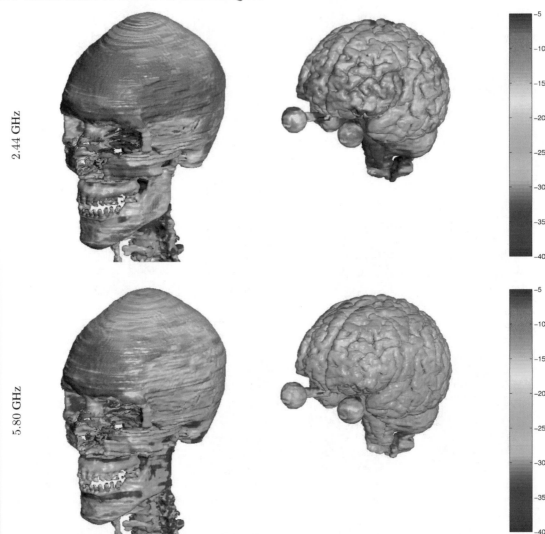

FIGURE 4.25: Normalized SAR distribution on the surface of regions associated with the bones, eyes, and the brain of human head for an exposure to a 2.44- and a 5.80-GHz electromagnetic plane wave with an incidence angle of $\varphi = 90°$.

maximum SAR values (Figure 4.26a) are observed at $-90°$ and $-20°$ on the skin and are at least four times larger than those of the eyes, which are proven more intense at $-30°$. Conversely, for 5.80 GHz the maximum values (Figure 4.26b) are accomplished at $0°$, with the case of the eyes being almost eight times smaller. Inspecting Figure 4.26c, one may easily come up with the conclusion that the

TABLE 4.5: Maximum values of diverse SAR indicators for different frequencies and power densities

INCIDENCE ANGLE (DEG)	2.44 GHz			5.80 GHz		
	MAX SAR	MAX SAR_{1g}	MAX SAR_{10g}	MAX SAR	MAX SAR_{1g}	MAX SAR_{10g}
−90	4.385	1.110	0.345	17.175	4.941	0.369
−75	3.910	1.575	0.387	18.353	5.882	0.330
−60	4.092	1.906	0.413	16.367	5.685	0.344
−45	4.151	2.077	0.504	19.227	4.501	0.417
−30	4.240	2.233	0.535	20.642	4.348	0.395
−15	4.275	2.215	0.495	22.894	4.307	0.397
0	3.817	1.949	0.488	24.220	4.566	0.421
15	3.926	1.575	0.471	19.581	4.081	0.397
30	3.714	1.510	0.424	16.309	3.752	0.357
45	3.455	1.511	0.416	18.419	3.616	0.358
60	3.409	1.742	0.401	18.859	3.888	0.366
75	2.903	1.208	0.338	16.660	4.272	0.367
90	2.585	0.909	0.274	15.067	4.067	0.338

maximum SAR_{1g} values for 2.44 GHz violate the IEEE threshold (1.6 W/kg), mainly in the angle range from −74° to 13°, whereas those in the eyes and brain are up to nine times below the limit.

In particular, power absorption in the brain is found to be virtually independent to the angle of incidence. On the other hand, and as deduced from Figure 4.26d, the maximum values of the SAR_{1g} indicator for 5.80 GHz violate the IEEE limit (about 2.2–3.7 times beyond) [48] at all angles of illumination. Such a behavior is not encountered in the case of the eyes and brain, where the safety distance is deemed fairly sufficient. Lastly, for the ICNIRP standard [49], all SAR_{10g} values

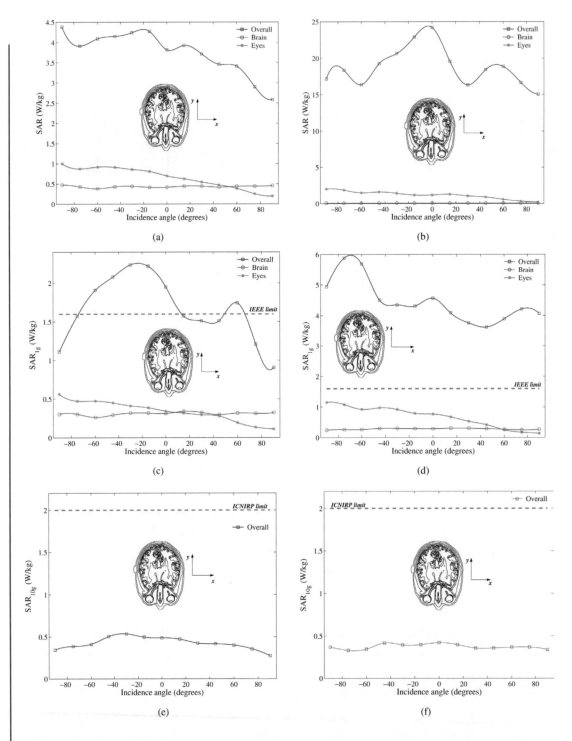

FIGURE 4.26: Maximum values of SAR, SAR_{1g}, and SAR_{10g} indicators in the coronal section of the human head versus angle of incidence at (a, c, e) 2.44 GHz and (b, d, f) 5.80 GHz.

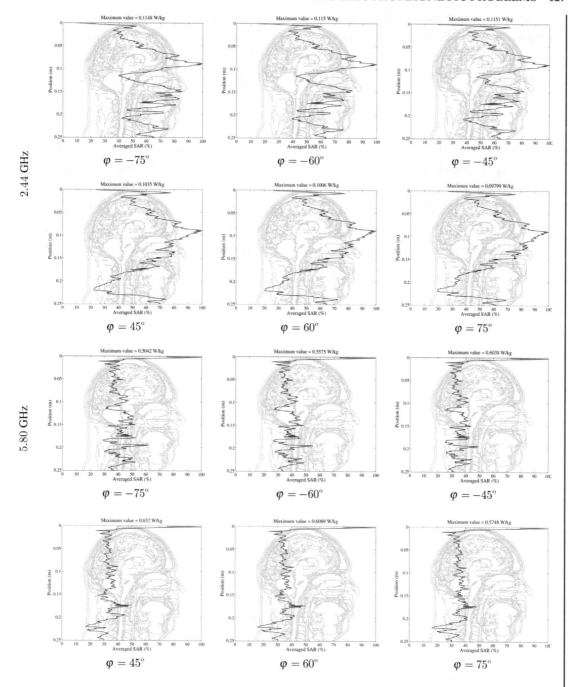

FIGURE 4.27: 2.44 and 5.80 GHz mean values (normalized to the maximum of each case) of SAR in a sagittal section of the human head for exposure at an electromagnetic plane wave with various incidence angles φ.

are found to be considerably lower than the maximum allowable limit (2 W/kg), as derived from Figure 4.26e and f for 2.44 and 5.80 GHz. Analogous remarks are extracted from Figure 4.27 that provides the mean SAR values in the sagittal section of the head for different φ. Here, maximization is detected at the level of the eyes (2.44GHz) and the top of the head (5.80 GHz).

4.4.2   Thermal Analysis

This paragraph is devoted to the thermal effects of WLAN configurations in the case of far-field human exposure to electromagnetic plane waves. Therefore, Figure 4.28 shows the maximum temperature rise generated via an illumination whose power density is equal *to* the maximum allowable threshold of the IEEE safety standard [48]. Specifically for the frequency of 2.44 GHz (Figure 4.28a), it is concluded that the most significant variation of 0.747°C occurs at the incident angle of −23°, whereas in the eyes the temperature can increase up to 0.419°C at 0°. Furthermore, the corresponding results in the brain exhibit a relative independence to the angle of incidence, with their

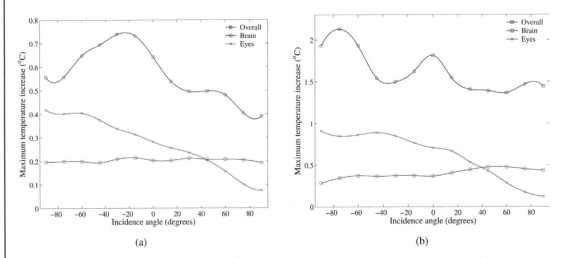

(a)   (b)

FIGURE 4.28: Maximum values of temperature rise for human exposure to an electromagnetic plane wave with a frequency of (a) 2.44 GHz and (b) 5.80 GHz. The maximum allowable equivalent power density is dictated by the corresponding IEEE safety standard for a monitored environment.

FIGURE 4.29: Temperature rise (normalized to the maximum value of each case) in two coronal sections for human exposure to a 2.44- and a 5.80-GHz electromagnetic plane wave propagating on the θ = 90° plane with different values of incidence angle φ.

2.44 GHz

$\varphi = -90°$ $\varphi = -60°$ $\varphi = -30°$ $\varphi = -15°$

$\varphi = 15°$ $\varphi = 30°$ $\varphi = 60°$ $\varphi = 90°$

5.80 GHz

$\varphi = -90°$ $\varphi = -60°$ $\varphi = -30°$ $\varphi = -15°$

$\varphi = 15°$ $\varphi = 30°$ $\varphi = 60°$ $\varphi = 90°$

maximum values fluctuating between 0.194°C and 0.215°C. In contrast, for the frequency of 5.80 GHz (Figure 4.28b), the maximum rise is observed at −75.5° and is equal to 2.13°C. Regarding the eyes, the temperature can rise up to 0.911°C, whereas for the brain computational investigation proves that this value is around 0.48°C.

The last set of results is given in Figure 4.29 which illustrates the normalized temperature rise at two coronal sections of the human head for an exposure at the $\theta = 90°$ plane and various φ incident angles. Again, the reduction of the radiation penetration depth owing to frequency augmentation is readily discernible.

4.5 ACCURACY AND IMPLEMENTATION ASPECTS

To validate the competences of the dispersive FDTD method, the computed average maximum SAR_{1g} and SAR_{10g} indicators are compared to their measured counterparts obtained from the bibliography. Both cases of cellular phone and WLAN radiation are taken into account, where in the former the device is located at four different "tilted" positions at $d = 2$ mm from the left ear, whereas in the latter far-field illumination is examined for four distinct incidence angles of the propagating wave. From the results of Tables 4.6 and 4.7, one can easily realize the high levels of accuracy

AVERAGE SAR TYPE	"TILTED" POSITION (DEG)	900 MHz		1800 MHz	
		MEASURED [17]	FDTD METHOD	MEASURED [17]	FDTD METHOD
1g	0	4.711	4.764	8.567	8.769
10g		3.101	3.178	4.482	4.565
1g	30	3.831	3.902	5.256	5.397
10g		2.378	2.430	3.229	3.302
1g	45	3.098	3.152	4.151	4.231
10g		1.937	1.983	2.631	2.688
1g	60	2.940	2.976	3.695	3.752
10g		1.541	1.581	2.067	2.097

TABLE 4.6: Comparison of measured and calculated maximum SAR values (cellular phones)

TABLE 4.7: Comparison of measured and calculated maximum SAR values (WLAN)

AVERAGE SAR TYPE	INCIDENCE ANGLE (DEG)	2.44 GHz		5.80 GHz	
		MEASURED [38]	FDTD METHOD	MEASURED [38]	FDTD METHOD
1g	0	1.925	1.949	4.653	4.566
10g		0.729	0.745	1.598	1.627
1g	30	1.485	1.510	3.736	3.752
10g		0.680	0.689	1.364	1.380
1g	45	1.483	1.511	3.599	3.616
10g		0.662	0.676	1.352	1.384
1g	60	1.708	1.742	3.847	3.888
10g		0.638	0.652	1.395	1.414

attained by the FDTD algorithm. In particular, the overall error varies between 1.1% and 2.6%; values that are deemed very satisfactory considering the large complexity and size of the EMC problem along with the multitude of different tissues.

Regarding the computational burden of the application, it should be noted that this depends heavily on the model of the human body or head, described in Section 4.2, and the number of tissues involved in the simulation. Generally, for the calculations performed in this chapter, memory requirements started from 650 MB to reach the level of 1.2 GB during the analysis of the most detailed discretizations. The corresponding CPU time (for an Intel© Pentium at 3.4 GHz with 3 GB of RAM) has been in the order of 3–5.5 h on condition that some adaptive nonuniform meshing has been used. However, in the case of the very fine human body model, the total CPU time approached the 20 h of time-domain simulation.

4.6 OTHER BIOMEDICAL APPLICATIONS

Aside from its hazardous influences, electromagnetic power radiated by specific antenna types may be successfully used for curing objectives inside the human body [46, 47]. There are three main classes of applications: *diagnostic imaging*, *thermal therapy*, and *medical implants*. Due to their multidisciplinary and sophisticated profile, the simulation of such problems opts for flexible and precise time-domain models without many nonphysical simplifications. To this goal, the FDTD method is again the most capable candidate, as readily deduced by the ample range of recent publications.

Concerning diagnostic imaging, it is important to state that remarkable accomplishments have been performed along the implementation at a clinical level of noninvasive ultra-wideband microwave imaging techniques for the detection of early-stage malignant tumors. In this work, the contribution of the FDTD method is chiefly spotted on the detailed characterization of the beam-forming radiator as well as on the proper discretization of all inhomogeneous tissue materials. Furthermore, analogous simulations have been conducted for the case of short-pulse transmitting/receiving bowtie or pyramidal horn antennas intended for cancer investigation. On the other hand, the frequency-dependent capabilities of the FDTD-related approaches are proven very instructive in magnetic resonance imaging, where the relationship between operating frequency and power dissipated inside the body constitutes a considerable request.

The category of thermal therapy precludes radio frequency and microwave hyperthermia which attempt to circulate the human body with a number of antennas that concentrate electromagnetic energy onto a superficially placed tumor [47]. This tissue heating process increases the temperature of the tumor beyond the normal level, thus immediately destroying its malignant parts or rendering it vulnerable to another simultaneous treatment of chemotherapy or radiotherapy. However, alternative concepts lead to the heating of deeply located tumors as well by means of annular phased antenna arrays, symmetrically arranged around the body. In this context, the role of the FDTD technique is the optimization of antenna magnitude and phase in conjunction with the correct position of the diverse radiators in the circumference of the patient.

Finally, the technology of medical implants seems to aggregate all intuitive ideas in antenna modeling for the construction of advanced devices for specialized purposes, such as cardiac peacemakers or bio-MEMS (microelectromechanical structures). Here, FDTD schemes lead to elaborate designs of miniaturized radiators with an emphasis on the profitable assimilation of the implant substances from the neighboring biological tissues.

REFERENCES

1. K. S. Kunz and R. J. Luebbers, *The Finite Difference Time Domain Method for Electromagnetics*. Boca Raton, FL: CRC Press, 1993.
2. A. Taflove and S. C. Hagness, *Computational Electrodynamics: The Finite-Difference Time-Domain Method*, 3rd ed. Norwood, MA: Artech House, 2005.
3. D. M. Sullivan, *Electromagnetic Simulation Using the FDTD Method*. Piscataway, NJ: IEEE Press, 2000.
4. H. Y. Chen and H. H. Wang, "Current and SAR induced in a human head model by the electromagnetic fields irradiated from a cellular phone," *IEEE Trans. Microwave Theory Tech.*, vol. 42, no. 12, pp. 2249–2254, Dec. 1994.

5. R. W. P. King, "Electromagnetic field generated in model of human head by simplified telephone transceiver," *Radio Sci.*, vol. 30, no. 1, pp. 267–281, 1995. doi:10.1029/94RS00510

6. M. Okoniewski and M. A. Stuchly, "A study of the handset antenna and human body interaction," *IEEE Trans. Microwave Theory Tech.*, vol. 44, no. 10, pp. 1855–1864, Oct. 1996. doi:10.1109/22.539944

7. K. V. Meier, R. Hombach, R. Kastle, R. Y. S. Tay, and N. Kuster, "The dependence of electromagnetic energy absorption upon human head modeling at 1800 MHz," *IEEE Trans. Microwave Theory. Tech.*, vol. 45, no. 11, pp. 2058–2062, Nov. 1997. doi:10.1109/22.644237

8. A. D. Tinniswood, C. M. Furse, and O. P. Gandhi, "Power deposition in the head and neck of an anatomically based human body for plane wave exposures," *Phys. Med. Biol.*, vol. 43, pp. 2361–2378, 1998.

9. K. R. Foster and L. S. Erdreich, "Thermal models for microwave hazards and their role in standards development," *Bioelectromagnetics*, vol. 20, pp. 52–63, 1999. doi:10.1002/(SICI)1521-186X(1999)20:4+<52::AID-BEM8>3.0.CO;2-7

10. A. Drossos, V. Santomaa, and N. Kuster, "The dependence of electromagnetic energy absorption upon human head tissue composition in the frequency range of 300–3000 MHz," *IEEE Trans. Microwave Theory Tech.*, vol. 48, pp. 1988–1995, Nov. 2000.

11. P. J. Dimbylow, "Fine resolution calculations of SAR in the human body fro frequencies up to 3 GHz," *Phys. Med. Biol.*, vol. 47, pp. 2835–2846, 2002. doi:10.1088/0031-9155/47/16/301

12. A.-K. Lee and J. K. Pack, "Study of the tissue volume for spatial-peak mass-averaged SAR evaluation," *IEEE Trans. Electromagn. Compat.*, vol. 44, pp. 404–408, May 2002.

13. M. C. González, A. Peratta, and D. Poljak, "Boundary element modeling of the realistic human body exposed to extremely-low-frequency (ELF) electric fields: Computational and geometrical aspects," *IEEE Trans. Electromagn. Compat.*, vol. 49, no. 1, pp. 153–162, Feb. 2007.

14. J. Toftgard, S. N. Hornsleth, and J. B. Andersen, "Effects on portable antennas of the presence of a person," *IEEE Trans. Antennas Propag.*, vol. 41, no. 6, pp. 739–746, Jun. 1993. doi:10.1109/8.250451

15. E. R. Adair, K. S. Mylacraine, and B. L. Cobb, "Human exposure to 2450 MHz CW energy at levels outside the IEEE c95.1 standard does not increase core temperature," *Bioelectromagnetics*, vol. 22, pp. 429–439, 2001. doi:10.1002/bem.70

16. P. Bernardi, M. Cavagnaro, S. Pisa, and E. Piuzzi, "SAR distribution and temperature increase in an anatomical model of the human eye exposed to the field radiated by the user antenna in a wireless LAN," *IEEE Trans. Microwave Theory Tech.*, vol. 46, pp. 2074–2082, Dec. 1998. doi:10.1109/22.739285

17. S. Schiavoni, P. Bertotto, G. Richiardi, and P. Bielli, "SAR generated by commercial cellular phones—Phone modeling, head modeling, and measurements," *IEEE Trans. Microwave Theory Tech.*, vol. 48, pp. 2064–2071, Nov. 2000.

18. P. Bernardi, M. Cavagnaro, S. Pisa, and E. Piuzzi, "Specific absorption rate and temperature increases in the head of a cellular-phone user," *IEEE Trans. Microwave Theory Tech.*, vol. 48, pp. 1118–1125, Jul. 2000. doi:10.1109/22.848494

19. T. V. Yioultsis, T. I. Kosmanis, E. P. Kosmidou, T. T. Zygiridis, N. V. Kantartzis, T. D. Xenos, and T. D. Tsiboukis, "A comparative study of the biological effects of various mobile phone and wireless LAN antennas," *IEEE Trans. Magn.*, vol. 38, no. 2, pp. 777–780, Mar. 2002. doi:10.1109/20.996201

20. P. Bernardi, M. Cavagnaro, S. Pisa, and E. Piuzzi, "Specific absorption rate and temperature elevation in a subject exposed in the far-field of radio-frequency sources operating in the 10–900-MHz range," *IEEE Trans. Biomed. Eng.*, vol. 50, pp. 295–304, Mar. 2003. doi:10.1109/TBME.2003.808809

21. M. Martinez-Burdalo, A. Martin, M. Anguiano, and R. Villar, "Comparison of FDTD calculated specific absorption rate in adults and children when using a mobile phone at 900 and 1800 MHz," *Phys. Med. Biol.*, vol. 49, pp. 345–354, 2004. doi:10.1088/0031-9155/49/2/011

22. S. Koulouridis and K. S. Nikita, "Study of the coupling between human head and cellular phone helical antennas," *IEEE Trans. Electromagn. Compat.*, vol. 46, no. 1, pp. 62–70, Feb. 2004. doi:10.1109/TEMC.2004.823612

23. K. Gosalia, J. Weiland, M. Humayun, and G. Lazzi, "Thermal elevation in the human eye and head due to the operation of a retinal prosthesis," *IEEE Trans. Biomed. Eng.*, vol. 51, pp. 1469–1477, Aug. 2004. doi:10.1109/TBME.2004.827548

24. O. Kivekäs, J. Ollikainen, T. Lehtiniemi, and P. Vainikainen, "Bandwidth, SAR, and efficiency of internal mobile phone antennas," *IEEE Trans. Electromagn. Compat.*, vol. 46, no. 1, pp. 71–86, Feb. 2004. doi:10.1109/TEMC.2004.823613

25. A. Pyrpasopoulou, V. Kotoula, A. Cheva, E. Nikolakaki, P. Hytiroglou, I. Magras, T. Xenos, T. Tsiboukis, and G. Karkavelas, "Bone morphogenic protein expression in newborn rat kidneys after prenatal exposure to radiofrequency radiation," *Bioelectromagnetics*, vol. 25, pp. 216–227, 2004.

26. J. Schuderer. T. Samaras, W. Oesch, D. Spät, and N. Kuster, "High peak SAR exposure unit with tight exposure and environmental control for in vitro experiments at 1800 MHz," *IEEE Trans. Microwave Theory Tech.*, vol. 52, pp. 2057–2066, Aug. 2004. doi:10.1109/TMTT.2004.832009

27. K. H. Chan, K. M. Chow, L. C. Fung, and S. W. Leung, "Effects of using conductive materials for SAR reduction in mobile phones," *Microwave Opt. Tech. Lett.*, vol. 44, pp. 140–144, Jan. 2005. doi:10.1002/mop.20569

28. J. Keshvari and S. Lang, "Comparison of radio frequency energy absorption in ear and eye region of children and adults at 900, 1800 and 2450 MHz," *Phys. Med. Biol.*, vol. 50, pp. 4355–4369, 2005. doi:10.1088/0031-9155/50/18/008

29. A. Hadjem, D. Lautru, C. Dale, M. F. Wong, V. F. Hanna, and J. Wiart, "Study of specific absorption rate (SAR) induced in two child head models and in adult heads using mobile phones," *IEEE Trans. Microwave Theory Tech.*, vol. 53, pp. 4–11, Jan. 2005.

30. A. Hirata, "Temperature increase in human eyes due to near-field and far-field exposures at 900 MHz, 1.5 GHz, and 1.9 GHz," *IEEE Trans. Electromagn. Compat.*, vol. 47, pp. 68–76, Feb. 2005. doi:10.1109/TEMC.2004.842113

31. J. Wang, M. Fujita, O. Fujiwara, K. Wake, and S. Watanabe, "Uncertainty evaluation of an *in vivo* near-field exposure setup for testing biological effects of cellular phones," *IEEE Trans. Electromagn. Compat.*, vol. 48, no. 3, pp. 545–551, Aug. 2006. doi:10.1109/TEMC.2006.877779

32. S. Ilvonen and J. Sarvas, "Magnetic-field-induced ELF currents in a human body by the use of a GSM phone," *IEEE Trans. Electromagn. Compat.*, vol. 49, no. 2, pp. 294–301, May 2007.

33. S. Mochizuki, H. Wakayanagi, T. Hamada, S. Watanabe, M. Taki, Y. Yamanaka, and H. Shirai, "Effects of ear shape and head size on simulated head exposure to cellular phone," *IEEE Trans. Electromagn. Compat.*, vol. 49, no. 3, pp. 512–518, Aug. 2007.

34. A.-K. Lee, H.-D. Choi, and J.-I. Choi, "Study on SARs in head models with different shapes by age using SAM model for mobile phone exposure at 835 MHz," *IEEE Trans. Electromagn. Compat.*, vol. 49, no. 2, pp. 302–312, May 2007.

35. T. Samaras, E. Kalampaliki, and J. N. Sahalos, "Influence of thermophysiological parameters on the calculations of temperature rise in the head of mobile phone users," *IEEE Trans. Electromagn. Compat.*, vol. 49, no. 4, pp. 936–939, Nov. 2007.

36. A. Hirata, M. Morita, and T. Shiozawa, "Temperature increase in the human head due to a dipole antenna at microwave frequencies," *IEEE Trans. Electromagn. Compat.*, vol. 45, pp. 109–116, Feb. 2003. doi:10.1109/TEMC.2002.808045

37. O. P. Gandhi and M. S. Lam, "An on-site dosimetry system for safety assessment of wireless base stations using spatial harmonic components," *IEEE Trans. Antennas Propagat.*, vol. 51, pp. 840–847, Apr. 2003. doi:10.1109/TAP.2003.809061

38. G. Kang and O. P. Gandhi, "Effect of dielectric properties on the peak 1- and 10-g SAR for 802.11 a/b/g frequencies 2.45 and 5.15 to 5.85 GHz," *IEEE Trans. Electromagn. Compat.*, vol. 46, no. 2, pp. 268–274, May 2004. doi:10.1109/TEMC.2004.826875

39. R. Cicchetti and A. Faraone, "Estimation of the peak power density in the vicinity of cellular and radio base station antennas," *IEEE Trans. Electromagn. Compat.*, vol. 46, no. 2, pp. 275–290, May 2004. doi:10.1109/TEMC.2004.826885

40. R. A. Abd-Alhameed, P. S. Excell, and M. A. Mangoud, "Computation of specific absorption rate in the human body due to base-station antennas using hybrid formulation," *IEEE Trans. Electromagn. Compat.*, vol. 47, no. 2, pp. 374–381, May 2005. doi:10.1109/TEMC.2005.847395

41. M. Fujimoto, A. Hirata, J. Wang, O. Fujiwara, and T. Shiozawa, "FDTD-derived correlation of maximum temperature increase and peak SAR in child and adult head models due to

dipole antenna," *IEEE Trans. Electromagn. Compat.*, vol. 48, no. 1, pp. 240–247, Feb. 2006. doi:10.1109/TEMC.2006.870816

42. R. Araneo and S. Celozzi, "Design of a microstrip antenna setup for bio-experiments on exposure to high-frequency electromagnetic field," *IEEE Trans. Electromagn. Compat.*, vol. 48, no. 4, pp. 792–804, Nov. 2006.

43. T. T. Zygiridis, E. P. Kosmidou, K. P. Prokopidis, N. V. Kantartzis, C. S. Antonopoulos, K. I. Petras, and T. D. Tsiboukis, "Numerical modeling of an indoor wireless environment for the performance evaluation of WLAN systems," *IEEE Trans. Magn.*, vol. 42, no. 4, pp. 839–842, Apr. 2006.

44. C. C. Davis, B. B. Beard, A. Tillman, J. Rzasa, E. Merideth, and Q. Balzano, "International intercomparison of specific absorption rates in a flat phantom in the near-field of dipole antennas," *IEEE Trans. Electromagn. Compat.*, vol. 48, no. 3, pp. 579–588, Aug. 2006.

45. C. Buccella, V. de Santis, and M. Feliziani, "Prediction of temperature increase in human eye due to RF sources," *IEEE Trans. Electromagn. Compat.*, vol. 49, no. 4, pp. 825–833, Nov. 2007.

46. E. J. Bond, X. Li, S. Hagness, and B. D. Van Deen, "Microwave imaging via space-time beamforming for early detection of breast cancer," *IEEE Trans. Antennas Propag.*, vol. 51, pp. 1690–1705, 2003. doi:10.1109/TAP.2003.815446

47. Special issue on "Medical applications and biological effects of RF/microwaves," G. Lazzi, O. P. Gandhi, and S. Ueno, Eds., *IEEE Trans. MicrowAVE. Theory Tech.*, vol. 52, pp. 1853–2083, 2004.

48. *IEEE standard for safety levels with respect to human exposure to radio frequency electromagnetic fields, 3 kHz to 300 GHz.* IEEE Standard C95.1-1999, New York: IEEE, 1999.

49. ICNIRP Guidelines, "Guidelines for limiting exposure to time-varying electric, magnetic, and electromagnetic fields (up to 300 GHz)," *Health Phys.*, vol. 74, 494–522, 1998.

50. *Council recommendation on 12 July 1999 on the limitation of exposure of the general public to electromagnetic fields (0 Hz to 300 GHz).* 1999/519/EC, *Official Journal of the European Communities,* 1999.

51. *Product standard to demonstrate the compliance of mobile phones with the basic restrictions related to human exposure to electromagnetic fields (300 MHz-3 GHz).* CENELEC Eur. Standard EN50360, Brussels, 2001.

52. *Basic standard for the calculation and measurement of electromagnetic field strength and SAR related to human exposure from radio base stations and fixed terminal stations for wireless telecommunication systems (110 MHz-40 GHz).* CENELEC Eur. Standard EN50383, Brussels, 2002.

53. *Limits on human exposure to radiofrequency electromagnetic fields in the frequency range from 3 kHz to 300 GHz.* Safety Code 6, Ministry of Health, Canada, 1999.

· · · · ·

CHAPTER 5

Time-Domain Characterization of EMC Test Facilities

5.1 INTRODUCTION

A critical issue for electromagnetic measurements is the provision of specialized conditions for both emission and immunity testing. Whether the goal of the study is for certification to a regulatory standard, user prerequisite, or troubleshooting, testing should be conducted in an environment that permits measurement to occur without disruption from external factors, such as ambient signals, weather conditions, outdoor noise interference, and ground-plane mutual coupling. For EMC compliance, in contrast, our concern focuses on the amount of electromagnetic field strength radiated by the structure under test, whereas for immunity, we focus on a certain energy portion that couples into the unit and causes performance degradation. Toward this direction, modern EMC test facilities [1, 2]—anechoic or semianechoic chambers [3–16], reverberation chambers [17–31], and transverse electromagnetic (TEM) cells [32–36]—have been developed as a basic "reflection-free" means of reliable testing.

It becomes evident that the fulfillment of contemporary specifications entails the improvement of active facilities and the design of novel ones. Nonetheless, this reconfiguration usually involves the replacement of their equipment or, under particular circumstances, the reconstruction of the entire arrangement. Taking into account the *excessive* building cost and the serious realization difficulties, precise *predictions* of a facility's performance, before its construction, are indeed crucial. A fairly competent tool for the completion of this objective is the implementation of *time-domain* numerical methods that lead to rigorous and informative solutions. Various efficient techniques have been so far proposed that thoroughly explore the properties of different media [4–6], create versatile computational models with distinct attractive features [7–9, 11, 14–16, 19, 29, 34], and launch robust approximation schemes for the computation of representative quantities [12, 13, 17, 18, 23, 27]. Among existing approaches, the FDTD algorithm has received a noteworthy attention for practical chamber evaluations. The driving motive of this popularity, apart from the technique's well-known merits, lies on its *flexibility* to manipulate diverse design perspectives and its successful *cooperation* with contemporary computer-based packages.

However, a major difficulty when numerically examining the wideband behavior of such electrically large EMC structures is the inhibiting system overhead. Not to mention the complications at low frequencies and the handling of dissimilar media interfaces, such as the absorber lining in anechoic chambers. The problem of *prolonged* simulations is more laborious for an arbitrarily curved equipment under test (EUT), where usual treatments lack to offer adequate outcomes. Lately, an unconditionally stable FDTD rendition has been presented via the alternating-direction implicit (ADI) process [37, 38]. By splitting each time-step into *two* parts, this approach removes the stability limit, theoretically allowing any possible choice for the simulation.

Bearing in mind the above observations, the present chapter provides a thorough time-domain characterization and optimal design of diverse EMC test facilities via the frequency-dependent FDTD, TLM, ADI-FDTD, hybrid FDTD/FETD, nonstandard FDTD, and PSTD methods. Moreover, special emphasis is drawn to a generalized curvilinear and unconditionally stable algorithm, implemented with robust 3-D dispersion-optimized schemes that accomplish an extensive decrease of lattice errors and enable the use of time-steps *significantly* beyond the Courant limit, as opposed to the conventional ADI approach.

5.2   FORMULATION OF THE ADI-FDTD METHOD

Having already described the implementation details of most time-domain techniques, in the previous chapters, this section deals with the establishment of the ADI-FDTD method both in its ordinary and nonstandard variant. Actually, the incorporation of this semi-implicit time-update concept allows for the selection of temporal increments far above the Courant condition, without sacrificing any other modeling asset.

5.2.1   Conventional Algorithm

Consistent with the initial idea, the discretized field components are staggered in space like in a typical grid, but collocated rather than staggered in time. Thus, for the solution evolving from time-step n to $n+1$, the FDTD iteration is separated into two sub-iterations: the first to advance from n to $n+1/2$ and the second to advance from $n+1/2$ to $n+1$ [37, 38]. Owing to their completely dual formulation, regarding the computation of the unknown quantities, our analysis will focus only on the description of the former.

Let us consider the x-directed part of Ampère's law, which for the first half-step can be discretized as

$$\varepsilon \left.\frac{\partial E_x}{\partial t}\right|^{n+1/4} + \left.\sigma E_x\right|^{n+1/4} = \left.\frac{\partial H_z}{\partial y}\right|^{n+1/2} - \left.\frac{\partial H_y}{\partial z}\right|^{n} \qquad (5.1)$$

Observe that the spatial derivatives of magnetic fields are calculated at two different time-steps, giving the average center point of $n + 1/4$. Alternatively, $\partial H_z/\partial y$ is substituted with a semi-implicit difference approximation of its so far unknown pivotal values at $n + 1/2$, whereas $\partial H_y/\partial z$ is evaluated explicitly from known field data at n. Via a fixed cell origin, (5.1) gives

$$\frac{2\varepsilon_{i,j,k}}{\Delta t}\left[E_x|_{i,j,k}^{n+1/2} - E_x|_{i,j,k}^{n}\right] + \frac{\sigma_{i,j,k}}{2}\left[E_x|_{i,j,k}^{n+1/2} + E_x|_{i,j,k}^{n}\right]$$

$$= \frac{1}{\Delta y}\left[H_z|_{i,j+1/2,k}^{n+1/2} - H_z|_{i,j-1/2,k}^{n+1/2}\right] - \frac{1}{\Delta z}\left[H_y|_{i,j,k+1/2}^{n} - H_y|_{i,j,k-1/2}^{n}\right], \tag{5.2}$$

where the unknown E_x quantity at $n + 1/4$, related to the losses term, has been replaced by the temporal average of its already obtained counterparts at n and $n + 1/2$. Similar formulae to (5.2) are also derived for the remaining E_y and E_z components. In fact, the key aim is to find the unknown electric components at time-step $n + 1/2$. Nonetheless, there are also unknown magnetic terms on the right-hand side of (5.2). To discard these quantities, the z-directed part of Faraday's law at $n + 1/4$ yields

$$\frac{2\mu_{i,j,k}}{\Delta t}\left[H_z|_{i,j,k}^{n+1/2} - H_z|_{i,j,k}^{n}\right] + \frac{\rho'_{i,j,k}}{2}\left[H_z|_{i,j,k}^{n+1/2} + H_z|_{i,j,k}^{n}\right]$$

$$= \frac{1}{\Delta y}\left[E_x|_{i,j+1/2,k}^{n+1/2} - E_x|_{i,j-1/2,k}^{n+1/2}\right] - \frac{1}{\Delta x}\left[E_y|_{i+1/2,j,k}^{n} - E_y|_{i-1/2,j,k}^{n}\right] \tag{5.3}$$

Assume, again, the update of E_x in (5.2) and use (5.3) to eliminate the unknown H_z terms at $n + 1/2$. In this way, the ensuing equation is extracted

$$\kappa_1 E_x|_{i,j,k}^{n+1/2} - \kappa_2 E_x|_{i,j+1,k}^{n+1/2} - \kappa_3 E_x|_{i,j-1,k}^{n+1/2} = \kappa_4 E_x|_{i,j,k}^{n} + \kappa_5 H_z|_{i,j+1/2,k}^{n} - \kappa_6 H_z|_{i,j-1/2,k}^{n}$$

$$- \kappa_7\left[H_y|_{i,j,k+1/2}^{n} - H_y|_{i,j,k-1/2}^{n}\right] - \kappa_8\left[E_y|_{i+1/2,j+1/2,k}^{n} - E_y|_{i-1/2,j+1/2,k}^{n}\right]$$

$$+ \kappa_9\left[E_y|_{i+1/2,j-1/2,k}^{n} - E_y|_{i-1/2,j-1/2,k}^{n}\right] \tag{5.4}$$

where

$$\kappa_1 = 1 + \left(\frac{2\Delta t}{\Delta y}\right)^2 p_{i,j,k}\left(q_{i,j+1/2,k} - q_{i,j-1/2,k}\right), \quad \kappa_2 = \left(\frac{2\Delta t}{\Delta y}\right)^2 p_{i,j,k}q_{i,j+1/2,k}, \quad \kappa_3 = \left(\frac{2\Delta t}{\Delta y}\right)^2 p_{i,j,k}q_{i,j-1/2,k}$$

$$\kappa_4 = p_{i,j,k}\left(4\varepsilon_{i,j,k} - \sigma_{i,j,k}\Delta t\right), \quad \kappa_5 = \frac{2\Delta t}{\Delta y}p_{i,j,k}q_{i,j+1/2,k}r_{i,j+1/2,k},$$

$$\kappa_6 = \frac{2\Delta t}{\Delta y} p_{i,j,k} q_{i,j-1/2,k} r_{i,j-1/2,k}, \ \kappa_7 = \frac{2\Delta t}{\left(4\varepsilon_{i,j,k} + \sigma_{i,j,k}\Delta t\right)\Delta z} p_{i,j,k},$$

$$\kappa_8 = \frac{(2\Delta t)^2}{\Delta x \Delta y} p_{i,j,k} q_{i,j+1/2,k}, \ \kappa_9 = \frac{(2\Delta t)^2}{\Delta x \Delta y} p_{i,j,k} q_{i,j-1/2,k},$$

with

$$p_{i,j,k} = \left(4\varepsilon_{i,j,k} + \sigma_{i,j,k}\Delta t\right)^{-1}, \quad q_{i,j,k} = \left(4\mu_{i,j,k} + \rho'_{i,j,k}\Delta t\right)^{-1}, \quad r_{i,j,k} = 4\mu_{i,j,k} - \rho'_{i,j,k}\Delta t$$

It is obvious that the repeated application of (5.4) to every j-coordinate along a y-directed line through the mesh, leads to a system of simultaneous equations for E_x, whose matrix is tridiagonal and easily manipulated by available numerical algorithms. Analogous outcomes are accomplished for the other two **E**-field components. If all electric quantities are determined, the first sub-iteration is completed with the computation of H_x, H_y, and H_z components at $n + 1/2$ in a fully explicit and direct regime.

As for the second sub-iteration from time-step $n + 1/2$ to $n + 1$, the role of the unknown terms in (5.1) and (5.2) is dually reversed. Hence, now, $\partial H_z/\partial y$ is treated explicitly from known values and $\partial H_y/\partial z$ receives the semi-implicit evaluation via the unknown pivotal field data at $n + 1$, whereas the construction of the tridiagonal system is equivalent to that of the first sub-iteration after the corresponding removal of the H_y terms. The advantage of the aforesaid procedure is easily recognized in 3-D EMC problems, where only two alternations are needed and numerical calculations are conducted along all coordinate axes.

5.2.2 Unconditionally Stable Nonstandard Schemes

Since its initial advent, the ADI-FDTD technique became the subject of an ongoing research that revealed its principal advantages and weaknesses [39–46]. Actually, this examination indicated that the stability limit can not be arbitrarily surpassed because dispersion and anisotropy errors are constantly increasing. Such artificial mechanisms, together with some nontrivial asymmetries, depend closely on grid resolution and spatial sampling rate, whereas their negative impact is radically amplified by the dispersive nature of the test facilities' lossy media and the necessity for broadband evaluations.

To ameliorate the prior defects, a 3-D curvilinear ADI-FDTD method whose time increments may greatly surpass the Courant condition without detrimental dispersion errors, is introduced. Moreover, given that the constitutive parameters, $\varepsilon = \varepsilon_0\varepsilon_r$ and $\mu = \mu_0\mu_r$, of most anechoic chamber materials (e.g., urethane tapered or ferrite tile absorbers) depend strictly on frequency, wideband efficiency estimations involving a range from 30 MHz to 3 GHz entail that the technique must be devised in a frequency-dependent fashion. For the achievement of this dependence, an

*N*th-order scheme with advanced spatial/temporal nonstandard operators, which encompasses the merits of the first-order Debye differential model [43, 44], is developed. In this manner, the undue complexity of the convolution integral approach is avoided, while structures that would otherwise call for nonuniform meshing are now represented by the proper tessellation pattern.

Assume a homogeneous, isotropic, linear, and lossy medium whose electric frequency dependence is denoted by the variation of complex permittivity $\varepsilon(\omega)$, as

$$\bar{\varepsilon}(\omega) = \varepsilon_0 \varepsilon_r(\omega) = \varepsilon_0 \left(1 + \sum_{s=1}^{N} \frac{\gamma_s}{1 + j\omega\tau_s} \right), \tag{5.5}$$

where γ_s and τ_s are coefficients signifying the maximum attenuation and relaxation time of the material, respectively. Their values, for a prefixed N, are extracted from $\varepsilon_r(\omega)$ measurements [6, 11] at certain frequencies through any fast optimization algorithm. In fact, regular chamber absorbers need two or three terms for their accurate approximation. Hence, the electric flux density $\mathbf{D} = [D_u D_v D_w]^T$ is given by

$$\mathbf{D} = \bar{\varepsilon}(\omega)\mathbf{E} = \varepsilon_0 \mathbf{E} + \varepsilon_0 \sum_{s=1}^{N} \frac{\gamma_s}{1 + j\omega\tau_s} \mathbf{E} = \varepsilon_0 \mathbf{E} + \varepsilon_0 \sum_{s=1}^{N} \mathbf{P}_s, \tag{5.6}$$

with $E = [E_u E_v E_w]^T$ the electric field intensity at generalized coordinates (u, v, w), $\mathbf{P}_s = \gamma_s E/(1 + j\omega\tau_s)$ a set of N transitional electric polarization vectors, and superscript T denoting the matrix transpose. This relation, along with the two Maxwell curl equations, are now ready for discretization via the enhanced finite-difference approximators that achieve superior levels of accuracy and rates of convergence.

5.2.2.1 Nonstandard Curvilinear Spatial/Temporal Operators.
The prime concept of the 3-D dispersionless schemes stems from the analysis of numerical solutions by a class of discrete models that use extra grid points near the central node [46–48], unlike traditional techniques. Therefore, consider the previous (u, v, w) system together with its $g(u, v, w)$ metrics and presume that all derivatives are computed with an Mth-order accuracy. In this context, the parametric spatial and temporal nonstandard operators are defined as

$$\mathbf{L}_\xi \left[f\big|_{u,v,w}^t \right] = \frac{q_1}{4\Delta\xi} \left\{ R_{\xi,S}^M \left[f\big|_{u,v,w}^t \right] + \sum_{\eta=1}^{3} f\big|_{\xi\pm\eta\Delta\xi/2}^t \right\} \quad \text{for} \quad \xi \in (u,v,w) \tag{5.7}$$

$$\mathbf{T} \left[f\big|_{u,v,w}^t \right] = \frac{1}{C_T(\Delta t)} \left(f\big|_{u,v,w}^{t+\Delta t/2} - f\big|_{u,v,w}^{t-\Delta t/2} \right) - q_2 \frac{\partial^3 f}{\partial t^3}\bigg|_{u,v,w}^t, \tag{5.8}$$

where $\Delta\xi \in (\Delta u, \Delta v, \Delta w)$ is the space increment determined by $\mathbf{L}_\xi[.]$ and Δt is the time-step of the simulation. Quantities q_1, q_2 are polynomial functions of $\Delta\xi, \Delta t$, which improve derivative evaluation.

Their main role is to establish the pertinent association amid different lattice mappings as well as to monitor the effect of any additional node to the overall consistency. In particular, after some algebra, one obtains

$$q_1(\Delta\xi) = (\vartheta + 1)(\Delta\xi)^M - \vartheta(\Delta\xi)^{M-1} \quad \text{with} \quad \vartheta \in [0, 1/2], \tag{5.9}$$

$$q_2(\Delta t) = -(\rho + 1)(\Delta t)^{M-1} + (\rho - 1)(\Delta t)^M \quad \text{with} \quad \rho \in [0, 1] \tag{5.10}$$

In (5.7), the construction of the $R_{\xi,S}^M[.]$ multidirectional operator must be conducted very carefully. Essentially, its supplementary degree of freedom S specifies stencil size $S\Delta\xi$ along each mesh direction with a customary value of $S = 3$. A very convenient, although not unique, definition reads

$$R_{\xi,S}^M \left[f|_{u,v,w}^t \right] = \frac{g(u,v,w)}{C_S(k_\xi S\Delta\xi)} \sum_{m=1}^M U_m^\xi \left\{ \sum_{l=1}^L W_{m,l}^\xi K_{\xi,l\Delta\xi}^{(m)} \left[f|_{u,v,w}^t \right] \right\} \tag{5.11}$$

Coefficients U_m and $W_{m,l}$ advance the modeling competences of the method in the treatment of geometric oddities or discontinuities by satisfying the next Mth-order gauges

$$\sum_{m=1}^M U_m^\xi = 1/2 \quad \text{and} \quad \sum_{l=1}^L W_{m,l}^\xi = 1 \,\forall m \tag{5.12}$$

Correction functions $C_T(\Delta t)$ and $C_S(k_\xi S\Delta\xi)$ of (5.8) and (5.11), respectively, allow the precise discretization of spatial and temporal derivatives in any frequency range. For wideband emission or immunity test procedures, these functions are specified by considering the frequency content and duration of the excitation. Thus, they subdue oscillatory or spurious modes that may corrupt the final waveform envelope for the illumination of the EUT [46]. In the case of low-frequency realizations (namely, 30–200 MHz), where strong antenna coupling occurs, this promising behavior is promptly retained. Even for a semianechoic facility where the virtually PEC floor is expected to spoil the outcomes due to serious grid reflections, correction functions are capable of orderly precluding all parasitic waves. Such an issue is deemed very important for chamber optimization, as numerically demonstrated below, because the representation of the excitation is crucial for resolving feasible designs. Here, $C_T(\Delta t)$ and $C_S(k_\xi S\Delta\xi)$ are described by the general combinations of

$$C_T(\Delta t) = \frac{e^{3\Delta t/2}}{\sqrt{1 - (3\Delta t/2)^2}} \cosh\left[4\Delta t \sqrt{1 - (3\Delta t/2)^2} \right], \tag{5.13}$$

$$C_S(k_\xi S\Delta\xi) = -\frac{e^p \Delta\xi}{2\sqrt{3p^2 - 1}} \cosh\left(4\Delta\xi \sqrt{p^2 - 1} \right) + e^p \sinh\left(4\Delta\xi \sqrt{p^2 - 1} \right), \tag{5.14}$$

with k_ξ the relevant k_u, k_v, or k_w component of the 3-D wavenumber vector $\mathbf{k}$ and $p = k_\xi S\Delta\xi/2$. The proper argument $k_\xi S\Delta\xi$ in (5.14) is acquired via the Fourier transform of certain electric and magnetic components at predetermined lattice positions that exhibit abrupt field variations. For instance, points near sharp pyramidal absorbers or the terminating border of a wedge lining are some characteristic details that can satisfactorily dictate the argument choice. After the required frequency range has been estimated, the spectrum's mean value is used to find the optimal $k_\xi S\Delta\xi$ that is capable of following every waveform change.

To finish with the definitions in (5.11), operators $K^{(m)}_{\xi,l\Delta\xi}[.]$ should be described. Targeting at the inclusion of most (if not all) available nodal setups, our intention is to enhance the weak two-point derivative approximation of the FDTD-based approaches. Hence, a w-directed $K^{(m)}_{w,l\Delta w}[.]$, for $S = 3$, is written as

$$K^{(m)}_{w,l\Delta w}\left[f|^t_{u,v,w}\right] = \frac{(\Delta w)^m}{5m-1} \sum_{r=-1}^{+1} \left[f|^t_{u-\Delta u,v+r\Delta v,w+rl\Delta w/2} - f|^t_{u+\Delta u,v-r\Delta v,w-rl\Delta w/2}\right] \quad (5.15)$$

where r assists the robust manipulation of regions near chamber's outer walls or composite absorber interfaces, i.e., cases where stencils extend at least two nodes on each side of a grid point. Observing the aforementioned algorithm, it should be stressed that (5.7), (5.8), (5.11), and (5.15) introduce a family of new 3-D higher-order nonstandard FDTD forms with optimal dispersion-reduction capabilities [46]. As a matter of fact, it is the topological consistency of (5.11) that overwhelms conventional counterparts which have been up to now used. Analytically, traditional operators are still prone to discretization shortcomings induced during the modeling of electrically large structures. Moreover, their limited number of nodes, despite the fine mesh resolution they may use, does not suffice for the correct depiction of geometrically demanding absorbers, thus yielding serious inaccuracies. Conversely, nonstandard finite-difference approximators have lately established a noteworthy potential in modern computational electromagnetics. Provided that lengthy propagation distances are essentially associated to multiple interactions from composite walls, the extraction of stable update formulae has a critical effect on the overall process. This stipulation is fulfilled by (5.7)–(5.15) which, owing to their notable accuracy and dispersion suppression, will be the fundamental tool for the derivation of the ADI-FDTD technique and its application to the study of EMC test facilities.

Integrating the theoretical background, so formulated, Maxwell's curl equations become

$$\text{Ampère's law: } \bar{\bar{G}}^H \mathcal{L}\left[\mathbf{H}\right] = \mathbf{T}\left[\mathbf{D}\right] + \sigma\mathbf{E} + \mathbf{J}, \quad (5.16)$$

$$\text{Faraday's law: } \bar{\bar{G}}^E \mathcal{L}\left[\mathbf{E}\right] = -\mu\mathbf{T}\left[\mathbf{H}\right], \quad (5.17)$$

$$\mathbf{D} = \varepsilon_0 \mathbf{E} + \varepsilon_0 \sum_{s=1}^{N} \mathbf{P}_s, \quad \mathbf{P}_s + \tau_s \mathbf{T}[\mathbf{P}_s] = \gamma_s \mathbf{E} \quad \text{for } s = 1, 2, \ldots, N, \tag{5.18}$$

with $\mathbf{H} = [H_u H_v H_w]^T$ the magnetic intensity, $\mathbf{J} = [J_u J_v J_w]^T$ the electric current density, σ the losses of the medium and $\overline{\overline{G}}^{\mathrm{H}}$, $\overline{\overline{G}}^{\mathrm{E}}$ 3×3 dual tensors whose elements, expressed in terms of $g(u, v, w)$, characterize the coordinate system of the problem. Note that (5.18) shows the *auxiliary* differential equations that relate the frequency-dependent $\mathbf{D}$, $\mathbf{E}$ vectors and are derived via the inverse Fourier transform of $\mathbf{P}_s$, assuming an $e^{j\omega t}$ variation. On the other hand, $\mathcal{L}[.]$ is the nonorthogonal curl operator given, along with (5.7), by

$$\mathcal{L}[.] = \begin{bmatrix} 0 & -\mathbf{L}_w & \mathbf{L}_v \\ \mathbf{L}_w & 0 & -\mathbf{L}_u \\ -\mathbf{L}_v & \mathbf{L}_u & 0 \end{bmatrix} \tag{5.19}$$

5.2.2.2 The Generalized ADI-FDTD Method.

Maintaining the idea of time-step division, the algorithm develops, again, the same two sub-iterations of its conventional variant. An important issue, however, is the substitution of electric current density components, encountered in (5.16), with an adjusted set of temporal averages, due to the inability to have their direct values at the intermediate time-steps. Hence, at $n + 1/2$, one gets

$$\sigma \mathbf{E}^{n+1/2} = 0.25\sigma \left[\left(\mathbf{E}^{n+1/2} + \mathbf{E}^n \right) + \left(\mathbf{E}^{n+1} + \mathbf{E}^{n+1/2} \right) \right]$$

$$\mathbf{J}^{n+1/2} = 0.5 \left(\mathbf{J}^{n+1/4} + \mathbf{J}^{n+3/4} \right), \tag{5.20}$$

where the $0.25\sigma(\mathbf{E}^{n+1/2} + \mathbf{E}^n)$, $0.5\mathbf{J}^{n+1/4}$ terms are used during the first and the $0.25\sigma(\mathbf{E}^{n+1} + \mathbf{E}^{n+1/2})$, $0.5\mathbf{J}^{n+3/4}$ terms during the second sub-iteration. As a consequence, in the interval between n and $n + 1/2$, the E_w component can be initially expressed by means of the w-directed part of (5.16) as

$$\mathbf{T} \left[D_w |_A^{n+1/2} \right] + 0.25\sigma \left(E_w |_A^{n+1/2} + E_w |_A^n \right) + 0.5 J_w |_A^{n+1/4} = g_{vw} \mathbf{L}_u \left[H_v |_A^{n+1/2} \right] - g_{wu} \mathbf{L}_v \left[H_u |_A^n \right], \tag{5.21}$$

for $A = (i, j, k + 1/2)$. The analysis of operator $\mathbf{T}[.]$, through (5.8), yields

$$D_w |_A^{n+1/2} = D_w |_A^n + C_T(\Delta t) \left\{ g_{vw} \mathbf{L}_u \left[H_v |_A^{n+1/2} \right] - g_{wu} \mathbf{L}_v \left[H_u |_A^n \right] \right.$$

$$\left. - 0.25\sigma \left(E_w |_A^{n+1/2} + E_w |_A^n \right) - 0.5 J_w |_A^{n+1/4} + \Gamma_A^{\mathrm{HO}} \right\}, \tag{5.22}$$

where Γ_A^{HO} is a function that aggregates all higher-order temporal differentiations of D_w conducted at n or earlier time-steps [14, 15]. As detected from (5.22), $\mathbf{L}_u[H_v]$ should be implicitly calculated, because it involves only unknown H_v pivotal $(i \pm 1/2, j, k + 1/2)$ values at $n + 1/2$, whereas its $\mathbf{L}_v[H_u]$

constituent is explicitly defined by the hitherto evaluated H_u quantities at n. To shed H_v, the u-directed part of (5.17) leads to

$$\mu \mathbf{T}\left[H_v|_B^{n+1/2}\right] = g_{vw}\mathbf{L}_u\left[E_w|_B^{n+1/2}\right] - g_{uv}\mathbf{L}_w\left[E_u|_B^n\right], \qquad (5.23)$$

for $B = (i \pm 1/2, j, k + 1/2)$. Expanding $\mathbf{T}[.]$, (5.23) becomes

$$H_v|_B^{n+1/2} = H_v|_B^n + \frac{C_T(\Delta t)}{\mu}\left\{g_{vw}\mathbf{L}_u\left[E_w|_B^{n+1/2}\right] - g_{uv}\mathbf{L}_w\left[E_u|_B^n\right] + \Gamma_B^{HO}\right\}, \qquad (5.24)$$

where Γ_B^{HO} is the function of higher-order H_v temporal derivatives. Nevertheless, (5.22) contains an additional unknown accredited to frequency dependence, namely, the D_w at $n + 1/2$. This term may be effectively handled by an unconditionally stable Crank-Nicolson scheme applied to (5.18). Thus, for $s = 1, 2, \ldots, N$, the sth differential equation is discretized as

$$P_{w,s}|_A^{n+1/2} = \left[\frac{\gamma_s C_T(\Delta t)}{4\tau_s + C_T(\Delta t)}\right]\left(E_w|_A^{n+1/2} + E_w|_A^n\right) + \left[\frac{4\tau_s - C_T(\Delta t)}{4\tau_s + C_T(\Delta t)}\right]P_{w,s}|_A^n + \frac{\Gamma_C^{HO}}{4\tau_s + C_T(\Delta t)}, \qquad (5.25)$$

with Γ_C^{HO} the appropriate higher-order function. In this manner, D_w is acquired via

$$D_w|_A^{n+1/2} = \varepsilon_0 E_w|_A^{n+1/2} + \varepsilon_0 \sum_{s=1}^{N} P_{w,s}|_A^{n+1/2}, \qquad (5.26)$$

while E_w at $n + 1/2$, after plugging (5.25) and (5.26) into (5.22), by

$$\alpha_1 E_w|_{i+1,j,k+1/2}^{n+1/2} - \alpha_2 E_w|_{i+1,j,k+1/2}^{n+1/2} - \alpha_3 E_w|_{i-1,j,k+1/2}^{n+1/2} = \alpha_4 E_w|_{i,j,k+1/2}^n + \alpha_5 J_w|_{i,j,k+1/2}^{n+1/4}$$
$$+ \alpha_6 \mathbf{L}_u\left[H_v|_{i,j,k+1/2}^n\right] - \alpha_7 \mathbf{L}_v\left[H_u|_{i,j,k+1/2}^n\right]$$
$$- \alpha_8\left\{\mathbf{L}_u\left[E_w|_{i+1/2,j,k+1/2}^n\right] + \mathbf{L}_w\left[E_u|_{i-1/2,j,k+1/2}^n\right]\right\} + \alpha_9 \sum_{s=1}^{N} P_{w,s}|_{i,j,k+1/2}^n, \qquad (5.27)$$

in which α_m ($m = 1, 2, \ldots, 9$) play the same role as κ parameters in (5.4). On the condition that α_m are constants, they can be promptly extracted at the beginning of the time-marching procedure [16]. Again, the application of (5.27) is repeated for each i along the u-mesh direction to provide the sparse three-band tridiagonal system of equations. Once E_w is obtained, the above technique is identically used for E_u and E_v, whereas magnetic and electric polarization vectors are explicitly found by the nonstandard time-domain relations.

Proceeding to the second sub-iteration from $n + 1/2$ to $n + 1$, the time update of E_w requires the reversal of roles amid $\mathbf{L}_u[H_v]$ and $\mathbf{L}_v[H_u]$, turning (5.21) to

$$\mathbf{T}\left[D_w|_A^{n+1/2}\right] + 0.25\sigma\left(E_w|_A^{n+1} + E_w|_A^{n+1/2}\right) + 0.5 J_w|_A^{n+3/4}$$
$$= g_{vw}\mathbf{L}_u\left[H_v|_A^{n+1/2}\right] - g_{wu}\mathbf{L}_v\left[H_u|_A^{n+1}\right] \qquad (5.28)$$

Now, the unknown quantities are H_u and D_w at time-step $n + 1$. Discarding these terms in a fashion similar to (5.23)–(5.26), the ensuing tridiagonal system of equations is constructed

$$\beta_1 E_w|_{i+1,j,k+1/2}^{n+1} - \beta_2 E_w|_{i+1,j,k+1/2}^{n+1} - \beta_3 E_w|_{i-1,j,k+1/2}^{n+1} = \beta_4 E_w|_{i,j,k+1/2}^{n+1/2} + \beta_5 J_w|_{i,j,k+1/2}^{n+3/4}$$

$$- \beta_6 \mathbf{L}_v \left[H_u|_{i,j,k+1/2}^{n+1/2} \right] + \beta_7 \mathbf{L}_u \left[H_v|_{i,j,k+1/2}^{n+1/2} \right]$$

$$+ \beta_8 \left\{ \mathbf{L}_w \left[E_u|_{i+1/2,j,k+1/2}^{n+1/2} \right] - \mathbf{L}_u \left[E_w|_{i-1/2,j,k+1/2}^{n+1/2} \right] \right\} + \beta_9 \sum_{s=1}^{N} P_{w,s}|_{i,j,k+1/2}^{n+1/2},$$

$$(5.29)$$

where β_m are the equivalents of α_m. Having computed all electric components, by solving (5.29), the remaining quantities are explicitly calculated and the process restarts for the next time-step.

A similar strategy holds for media such as ferrite tiles, whose permeability μ depends on frequency variations. In this situation, a constitutive expression, such as (5.18), is developed between s auxiliary magnetic polarization vectors and the magnetic flux density $\mathbf{B} = [B_u B_v B_w]^{\mathrm{T}}$. Note also that the extension of the algorithm to hybrid absorbers, involving both types of dispersive media, does not require any strenuous modifications, because each material is represented by the appropriate sum of interpolated measurement data associated with the variation of ε or μ. Finally, it should be noted that the computational burden of the generalized methodology is fairly confined, because of its competence to build practically dispersionless models and therefore allow for the implementation of coarse lattice resolutions at notably affordable simulation times.

5.3   EXPERIMENTAL AND MEASUREMENT PROCEDURES

The foremost objective of time-domain methods in this area of EMC analysis is the development of a mature and cost-effective design tool for fast and reliable test-facility performance evaluations before actual construction. In this manner, existing chambers that do not fulfill the recently ascertained specifications can be reequipped with up-to-date components, tailored to their initial dimensions. So, the structure's size is not seriously modified and the need for building a new facility is avoided with the benefit of substantial savings. Nevertheless, to validate the capabilities of these numerical techniques, their outcomes should be compared to several measurement and reference data, obtained by certain procedures according to international EMC immunity and emission standards. Consequently, this section is devoted to the brief description of these processes, and also provides instructive hints about some characteristic testing possibilities.

5.3.1   Open-Area Test Sites

An open-area test site (OATS) is the universally approved environment for conducting radiated emission certifications [1, 2, 49]. For its accurate operation, an OATS needs a calibrated receiving antenna, a ground-plane structure, and, most importantly, an amply remote distance from metallic objects and high-ambient fields. In fact, the use of OATS offers several advantages to researchers, ranging from the establishment of standardized measurement protocols to the opportunity of involving various radiating/transmitting components during any study. However, a primary disadvantage, aside from its large size and open-region character, is the obligation to search the entire frequency spectrum for unintentional radiated emissions within an electromagnetic environment, carrying white noise that surpasses the propagated signals from the device under test. Moreover, such an outsized area of land may be too expensive to procure in a densely populated industrial region. Not to mention that the ambient environment will be probably too high to ensure meaningful measurements. For locations where climate exhibits practical difficulties, the feasibility of an OATS must be carefully considered, because the cost for a covered site with material transparent to electromagnetic waves constitutes a substantial investment.

In particular, an OATS must be flat, free of overhead wires, and away from reflecting obstacles, where the EUT and antenna form the foci of an ellipse with a major diameter twice their distance, usually set at 3, 10, or 30 m, depending on the specifications and physical size of the test. For convenience, a sufficiently larger surrounding area without reflecting objects contains the control room and instrumentation. As an alternative, this infrastructure can be directly installed below the OATS, separated by a solid ground plane or using an available space in a reasonable distance from the range. On the other hand, the metallic ground plane is typically rectangular with a width twice the maximum test unit dimension. It should not have gaps or voids that comprise a significant fraction of a wavelength at around 1 GHz, whereas any inaccuracies aroused by radio frequency (RF) scattering from its edges must be minimized by terminating the ground plane into the nearby soil. To the above accessories, one should add the EUT turntable—encountered in all test facilities—which can rotate continuously and facilitate the determination of maximum radiation direction at each emission frequency. Lastly, the preferred antenna is the half-wave long telescopic dipole, placed on a remotely controlled positioner. By changing the height of the antenna, reflected signals from the ground plane can be discovered. The turntable is then rotated to determine the maximum radiation pattern of the propagated signal. After selecting the angle on the turntable that produces the highest levels of energy, the antenna obtains its definite location.

5.3.2   Performance Optimization and Basic EMC Quantities

Among the most important indicators for the testing competence of an anechoic or semianechoic chamber are the *field uniformity* and *normalized site attenuation (NSA)*. Both quantities are evaluated via a certain and systematically established process in consistency with the suitable international standards. Concerning the former indicator, it should be stressed that radiated immunity measurements are covered under the basic standard EN 61000-4-3 [50] and examine whether a piece of equipment functions correctly in the presence of an enforced electromagnetic field. Given that such tests refer to free space, they are basically accomplished in a fully anechoic chamber. Compatible with [50], the process for quantifying the effectiveness of the facility evaluates field uniformity that refers to the deviation of the field strength level across the face of the EUT opposite to the transmitting antenna during the test. It is measured by collecting data from a probe placed at the nodes of a 16-point, 1.5×1.5 m grid, located $d = 80$ cm above the turntable in a transverse or longitudinal layout, as presented in Figure 5.1a and 5.1b, for the frequency interval from 30 MHz to 1 GHz. If the computed values do not fluctuate beyond 6 dB at 12 out of the 16 positions (i.e., 75%), then the chamber is suitable for immunity testing, because it guarantees that the EUT is subjected to known field levels. In other words, the tolerance on the specified test field is 0 to +6 dB. A summary of the basic aspects regarding the measurement of field uniformity is given in Table 5.1. Note that EN 61000-4-3 currently does not have procedures for validating sites above 1 GHz. However, particular processes for smaller test planes or multiple transmit positions to consider the directive nature of the antennas at higher frequencies, are being developed [2].

On the other hand, the more difficult emission measurements investigate whether the electric-field strengths emitted by an apparatus are below a maximum threshold over a specific frequency range. They are primarily referenced to the respective values obtained in an OATS having a large PEC ground plane. It is to be emphasized that these values are not ideal. In fact, several researchers have demonstrated that their trace exhibits a ripple compared to an ideal OATS due to antenna

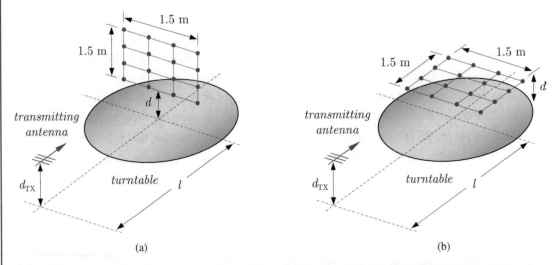

(a)   (b)

FIGURE 5.1: The 16-point grid for the evaluation of field uniformity: (a) transverse and (b) longitudinal layout.

TABLE 5.1: Experimental aspects for the measurement of field uniformity	
Measurement standard	EN 61000-4-3
Frequency	26 MHz–1 GHz
Distance l	1 or 3 m
Transmitting antenna height, d_{TX}	1.55 m
Receiving antenna height	1.55 m
Antenna calibration standard	ANSI C63.5
Field probe area	16 points on 1.5×1.5 m grid
Transmitting and receiving antenna	Biconical: 30–200 MHz
	Log-periodic: 200 MHz–1 GHz
	Double ridge horn: 1 GHz or higher

mismatch when the measurement or simulation is performed according to the typical procedure. Nevertheless, because such measured data are officially stated in ANSI C63.6 [51] and other national standards collectively provided in [1], they are always deemed very reliable. Therefore, herein, and for the purpose of our comparisons, the trace of these data is designated as "standard reference OATS". Due to its equivalence to the OATS function, the semianechoic chamber is used for this type of testing. The quantity to be specified is the NSA, which is measured according to the process established in ANSI C63.4 [52] and CISPR 22 [53] standards. More specifically, this quantity is defined as a function of the reference signal level, the spectrum analyzer reading, and the free space transmitting and receiving antenna factors, determined in terms of the ANSI C63.5 [54]. Based on the policy of these standards, if the normalized attenuation level of the site is within ±4 dB, for a volumetric measurement, then the structure is capable of conducting emission measurements without arousing artificial sources of error. The preceding measurement conditions for the NSA are jointly presented in Table 5.2.

A crucial aspect for the achievement of trustworthy measurements in any chamber is the calibration and position scanning of the transmitting and receiving devices, placed 1, 3, or 10 m away from the EUT depending on the selected standard. In our measurements, the antennas used for radiated field EMC testing have been proven of key importance as compared to the rest of the

TABLE 5.2: Experimental aspects for the measurement of normalized site attenuation	
Measurement standard	ANSI C63.4 and CISPR 22
Frequency	80 MHz–1 GHz
Distance l	1, 3, or 10 m
Transmitting antenna height, d_{TX}	1.5 m
Receiving antenna height	1.5 m
Antenna calibration standard	ANSI C63.5 and EN 50147-3
Transmitting and receiving antenna	Biconical: 30 MHz–200 MHz
	Log-periodic: 200 MHz–1 GHz

instrumentation. Actually, we have used many commercially available brands, such as the older biconical (range, 30–200 MHz) and log-periodic (range, 200–1000 MHz) designs as well as the more contemporary horn or hybrid antennas that can cover the whole frequency band from 26 MHz to 5 or 10 GHz in one sweep by combining and matching a biconical to a log-periodic section. Such composite antennas are often larger than a meter in width and length. Radiated immunity measurements require low voltage standing wave ratio and high gain so that the smallest possible amplifier can be used to generate the required field. Hence, such antennas are calibrated at 1- or 3-m distance to attain the optimal power with a given signal input as denoted by ANSI C63.5. Conversely, radiated emission testing opts for high-precision devices that are calibrated as pairs on a reference OATS according to ANSI C63.5 with their free space factors provided by EN 50147-3 [55]. All radiators are connected to the external environment via coaxial cables or optical fibers combined with the appropriate electro-optical converters. It is also stressed that the properties of the amplifiers and attenuators required for the optimal function of the anechoic or semianechoic chamber are defined by the ANSI C63.4 or the CISPR 22 standards.

5.3.3   Numerical Simulation Premises

Assessment of a test facility's efficiency opts for the frequency range—preferably the broader one—in which the structure provides an environment equivalent to that of free space, often called the quiet zone. Hence, the excitation scheme must be very carefully selected to ensure a rich frequency content from 30 MHz to 3 GHz or even more. Following the traditional practice, the device is handled like an external electric current density source launched in the computational domain by means of Ampère's law [56, 57]. This approach is deemed fairly adaptable, because it allows the choice of any excitation, anywhere in the lattice, with an arbitrary phase, amplitude, and polarization. In this work, we use a broadband Gaussian pulse with an exponential decay. For instance, the density of a current directed toward the u axis is expressed as

$$J_u|_{i,j,k}^n = I_u|_{i,j,k}^n / (\Delta v \Delta w) \quad \text{with} \quad I_u|_{i,j,k}^n = I_0 e^{-[(n-n_0)/n_d]^2}, \tag{5.30}$$

where I_0 is the amplitude and $n_0 \geq 3n_d$, n_d are coefficients that control the pulse content. Alternatively and for the sake of completeness, the first derivative of (5.30) has been also implemented.

It should be clarified that the FETD and nonstandard FDTD method are used in regions of fine geometric details, such as near the EUT, the corners of an anechoic chamber or the different parts of the mode stirrer of a reverberation chamber. For the remaining sections of the domain, the prior schemes are hybridized with the FDTD, the ADI-FDTD, or the TLM technique, whereas the PSTD algorithm is engaged for the case of periodic absorber linings or ferrite-tile arrangements. To this end, our analysis, also, focuses on media reflectivity $R_f = 20 \log_{10}(|E_r/E_i|)$, as an indicator of the absorber's wave annihilation rate in terms of the reflected (E_r) and the incident (E_i) electric fields. The lower the R_f, the *better* the function of the lining and in extent of the entire facility. Moreover, the deviation of the NSA from OATS (fulfillment of the ±1 dB criterion [51–53]) and electric-field temporal variation are studied especially for elongated simulations. Observe that in all numerical predictions, the unbounded discretized OATS is truncated via an eight-cell perfectly matched layer developed for nonorthogonal meshes [46]. Furthermore, the Courant-Friedrich-Levy number (CFLN) = $\Delta t / \Delta t_{\max}^{\text{md}}$ is defined for the ADI-FDTD schemes, with $\Delta t_{\max}^{\text{md}}$ the maximum temporal increment denoted via the stability criterion of the second-order (md = FDTD) or the nonstandard FDTD (md = ND) method, given by

$$\Delta t \leq \Delta t_{\max}^{\text{FDTD}} = \frac{1}{v} \left[\frac{1}{(\Delta x)^2} + \frac{1}{(\Delta y)^2} + \frac{1}{(\Delta z)^2} \right]^{-1/2}, \tag{5.31}$$

$$\Delta t \leq \Delta t_{\max}^{\text{ND}} = \frac{3}{v\pi} \sin^{-1}(0.7) \left(\sum_{\zeta=u}^{w} \sum_{\xi=u}^{w} \frac{1}{\Delta\zeta\Delta\xi} \right)^{-1/2}, \tag{5.32}$$

where v is the problem's phase velocity. Results are compared with measurement or reference data obtained by classical approaches to validate the formulation, even for large CFLNs.

5.4   ANECHOIC AND SEMIANECHOIC CHAMBERS

Anechoic chambers are metallic wall-shielded enclosures lined with low-reflection absorber arrays that isolate the internal region from ambient signals over a wide range of frequencies [1], as shown in the general depiction of Figure 5.2a and 5.2b. Because of ability to simulate free space to an acceptable degree, these structures are most frequently used for immunity investigations. In common practice, it is desirable to leave the floor reflective to act like the ground plane of an OATS but then cover the floor with removable absorber material to reduce resonant effects and ensure a uniform field required for radiated immunity testing.

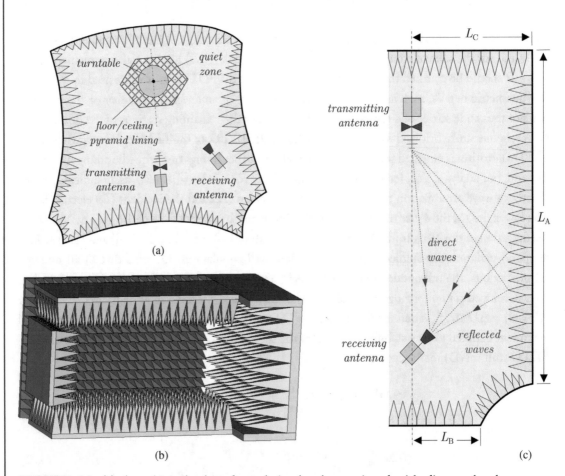

FIGURE 5.2: (a) An arbitrarily shaped anechoic chamber equipped with diverse absorber arrays. (b) Side view of an anechoic chamber computational model. (c) Top view of a partially curved semian-echoic chamber. The distance between the transmitting bilog and receiving horn antenna can vary in order to control the direction of the reflected waves.

On the other hand, semianechoic chambers (Figure 5.2c), with absorbers placed on at least a portion of their conducting floor, are used to a large extent for emission measurements.

5.4.1 Models of Different Absorber Linings

In general, the material used for absorber-lined chambers—either anechoic or semi-anechoic—is the carbon-loaded polyurethane foam in the shape of pyramidal cones or wedges in combination with ferrite tiles [1, 2, 6]. The selection of their basic geometric characteristic as well as the proper pattern used at the walls, floor, and ceiling of the chamber is of paramount importance, because these factors control the propagation directions of the dominant reflected waves and enable the construction of a reliable quiet zone. An example of a hybrid, general-shape absorber lining comprising curvilinear (specified by the arbitrary function f_{surf}) pyramids and backed by thin ferrite tiles is illustrated in Figure 5.3a. The presence of the optional air and lossy layers offers additional degrees of freedom for the tuning of the structure at specific frequencies. Similarly, Figure 5.3b presents an alternative tapered absorber, indicating the noteworthy design opportunities that can be verified by means of time-domain numerical methods able to model such complex realizations in due time.

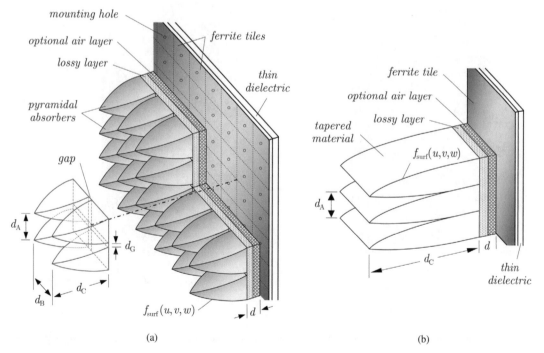

(a) (b)

FIGURE 5.3: (a) A hybrid urethane absorber array consisting of curvilinear pyramids and backed by thin ferrite tiles. (b) Detail of a tapered pyramidal absorber backed by thin ferrite tiles.

Aside from the preceding linings, Figure 5.4 shows two characteristic and commonly encountered absorber arrays consisting of 45° twisted pyramids and alternating wedges. Observe the independently chosen span angles, α, of the latter which, obviously, guarantee supplementary flexibility to the overall structure.

However, the efficiency of an absorber is not solely an issue of shape or proper material selection; it does depend on the distribution (layout) of each element in the lining that covers a wall. A very competent configuration whose absorbers follow a combined spherical/Chebyshev pattern, often found in the area behind the EUT, is displayed in Figure 5.5a. Moreover, special treatment is required at the two-wall edges and corners of the chamber, where absorbers meet and intersect one another. It should be noted that a nonoptimized setup at these locations is very likely to degrade the performance of the chamber no matter how carefully the other linings have been placed. To this goal, Figure 5.5a and b gives two popular array terminations, which are also used in the numerical simulations of this chapter.

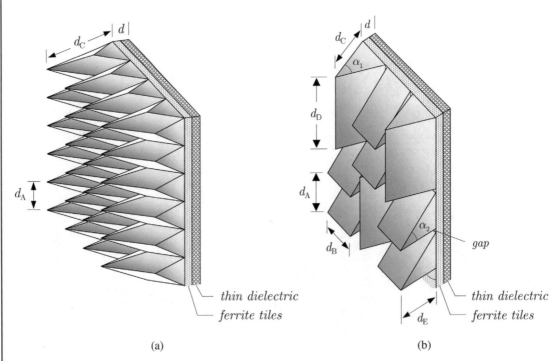

(a)   (b)

FIGURE 5.4: Typical arrays of urethane absorber linings comprising (a) 45° twisted pyramids and (b) alternating wedges.

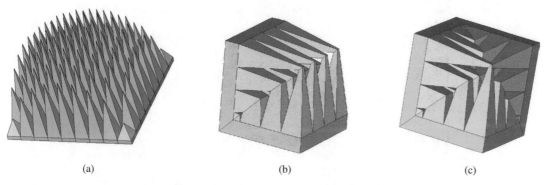

(a)   (b)   (c)

FIGURE 5.5: (a) An array of pyramidal absorbers at a combined spherical/Chebyshev layout and absorber arrangement at a (b) two-wall edge and (c) corner of a chamber.

5.4.2 Wideband Analysis of EMC Anechoic Chambers

Let us first address the inclined-wall dual reflector compact range anechoic facility, given in Figure 5.6, whose width is 7.4 m, $L_A = 6.2$ m, $L_B = 4.4$ m, $L_C = 5.4$ m, $L_D = 1.6$ m, $L_E = 7.8$ m, and $L_F = 11.3$m. Additional dimensions for the chamber's structural details are: $\theta_1 = 120°$, $\theta_2 = 7.5°$, $l_A = 3.2$ m, $l_B = 2.3$ m, $l_C = 2.2$ m, $l_D = 2.0$ m, $l_E = 1.3$ m, $l_F = 2.6$ m, and $l_G = 1.2$ m. As observed from the schematic depiction, the facility is lined with 45° twisted pyramids ($d_A = 0.41$ m, $d_C = 0.84 + 0.36 = 1.2$ m, $\varphi_1 = 150°$, $\varphi_2 = 100°$, 25% carbon-loaded), backed by 7.2-mm ferrite tiles and properly tuned ($\sigma = 2.34$ S/m) lossy layers as well as with alternating wedges ($d_A = 0.28$ m, $d_B = 0.32$ m, $d_D = 0.56$ m, $d_E = 1.2$ m, $d = 0.36$ m, $\alpha_1 = 30°$, and $\alpha_2 = 15°$). The field distribution, transmitted by high-gain corrugated horns along with the subreflector (diameter, 2.93 m), illuminates the main reflector (diameter, 4.2 m) which, in its turn, maps the propagating beam to a uniform plane wave front that illuminates the EUT. For the precise simulation of the very complicated preceding problem, the advanced ADI-FDTD method of Section 5.2.2 is hybridized with the FETD technique. The former scheme is anticipated to drastically reduce the overall simulation time, whereas the latter is expected to properly model all acute oddities around the interface of the two absorber linings, the two reflectors, and the boundaries of the quiet zone. Hence, after the calculation of the optimal γ_s, τ_s for $N = 3$, the space is discretized in $90 \times 88 \times 72$ cells with $\Delta t_{max}^{ND} = 0.203$ ns and 102,874 finite elements. In this context, the efficiency of the chamber is demonstrated in Figure 5.7, which provides field uniformity as the maximum electric-field deviation between 12 out of 16 points in the quiet zone versus frequency. Results indicate that only the hybrid algorithm can accurately follow the reference data of [1], whereas the staircase second-order one—for a 85% denser lattice with $\Delta t_{max}^{FDTD} = 0.059$ ns and small CFLN—is totally misleading. To underline the impact of dispersion as time-step increases beyond

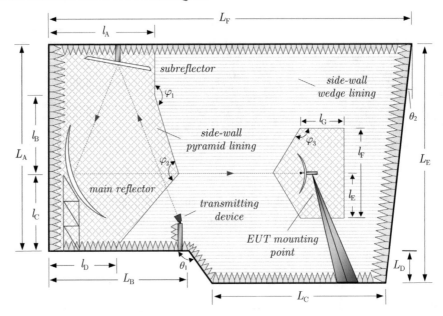

FIGURE 5.6: Basic elements and operation principle of an inclined-wall dual reflector compact range facility.

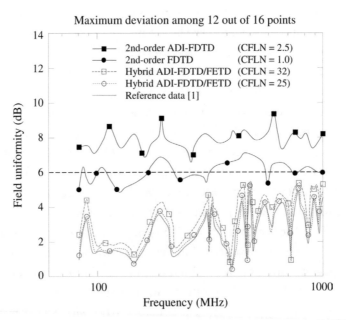

FIGURE 5.7: Field uniformity of a dual reflector range anechoic chamber equipped with 45° twisted pyramids.

Electric field (V/m)   Electric field (V/m)

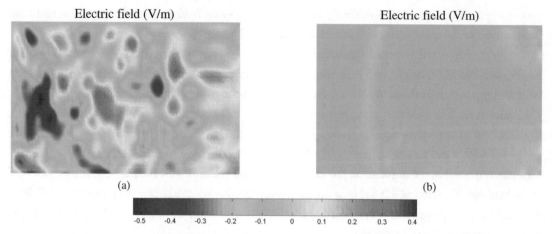

(a)   (b)

FIGURE 5.8: Electric-field amplitude through a rectangular area, 0.95 m above the turntable, in a dual reflector anechoic chamber at 80 MHz: (a) second-order ADI-FDTD method (CFLN = 2.47) and (b) hybrid ADI-FDTD/FETD method (CFLN = 26).

the Courant limit, electric field variation, through a rectangular area located 0.95 m above the turntable, is computed at 80 and 230 MHz by the second-order ADI-FDTD technique (Figures 5.8a and 5.9a for CFLN = 2.47, 65,000 time-steps) and the enhanced formulation (Figures 5.8b and 5.9b for CFLN = 26, 2950 time-steps). As expected, the detrimental dispersion errors spoil the former group of simulations via highly resonant modes. On the other hand, the latter group yields

Electric field (V/m)   Electric field (V/m)

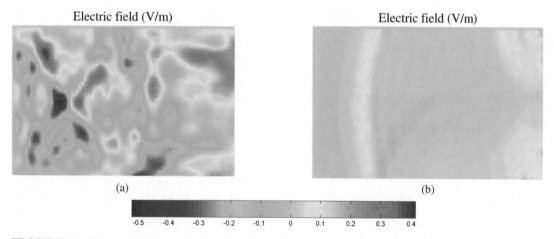

(a)   (b)

FIGURE 5.9: Electric-field amplitude through a rectangular area, 0.95 m above the turntable, in a dual reflector anechoic chamber at 230 MHz: (a) second-order ADI-FDTD method (CFLN = 2.47) and (b) hybrid ADI-FDTD/FETD method (CFLN = 26).

promising predictions, despite the large structural complexity, because it evaluates the actual field distribution in the quiet zone without arousing artificial patterns.

Next, the partially curved semianechoic chamber of Figure 5.2c with $L_A = 6.2$ m, $2L_B = 3.8$ m, $2L_C = 5.2$ m and a height of 5 m is examined. The curvilinear walls, behind the receiving antenna located 3 m away from the transmitting one, accomplish the correct structural adjustments for a reliable quiet zone. The facility is equipped with the *optimized* hybrid urethane absorbers ($d_A = 0.20$ m, $d_C + d = 0.46 + 0.15 = 0.61$ m, 27% carbon-loaded) illustrated in Figure 5.3b, backed by a 5.5-m-thick ferrite tile with a gap size of 0.2 mm and a 16.2-mm dielectric layer of $\varepsilon_r = 2.15$. For the frequency-dependent materials, we implement the ideas of (5.5) and select $N = 3$. Indicatively, the variation of $\varepsilon_r(\omega)$ between 30 MHz and 3 GHz is approximated by $\gamma_1 = 652.781$, $\tau_1 = 63.52$ ns, $\gamma_2 = 173.914$, $\tau_2 = 92.81$ ns, and $\gamma_3 = 854.257$, $\tau_3 = 56.74$ ns coefficients with a similar process holding for the relative magnetic permeability of ferrite tiles.

The structure is simulated by means of a combined ADI-FDTD/PSTD and a nonstandard (ND) FDTD/FETD that take into account the periodicity of the pyramidal linings and the abrupt field variations in the region of the EUT. Figure 5.10a and 5.10b gives the NSA for vertical (V) and horizontal (H) polarization, respectively, and different layouts. It should be mentioned that due to the advanced features of the latter method ($M = 3$, $L = 2$) the domain is divided in $76 \times 62 \times 78$ cells with $\Delta x = 0.102$ m, $\Delta y = 0.077$ m, $\Delta z = 0.096$ m, and 52,468 finite elements for the treatment of curved parts and the EUT in the quiet zone. Thus, grid resolution reaches the notable level of

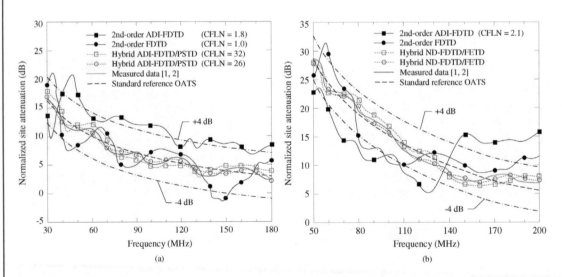

FIGURE 5.10: NSA for a 3-m partially curved semianechoic chamber. (a) Vertical and (b) horizontal polarization.

λ/7 (λ is the corresponding wavelength), thereby *avoiding* the excessive λ/150 choices of usual approximations. Based on measured data [1, 2], certifying the chamber's suitability for emission measurements, our aim is to investigate whether the combined techniques can predict such an issue. As deduced, this is achieved even for low frequencies without any dispersion errors. In contrast, the common ADI-FDTD and FDTD schemes fail to meet the ±4 dB limit, regardless of the 170 × 160 × 202 lattice with the far smaller Δt_{max}^{FDTD} = 0.061 ns and CFLN. The improved accuracy and savings (almost 90% grid and CPU time reduction) are also confirmed via Figure 5.11, presenting the deviation from OATS results for both types of polarization.

Proceeding to a more involved arrangement, the tapered anechoic chamber of Figure 5.12 is analyzed, whose shape forms an effective source antenna array leading to large frequency ranges (e.g., from 300 MHz to 16 GHz). In fact, extensive research has shown that this notable frequency coverage is attained if the taper angle does not exceed 36°. The dimensions of the 5.4-m-high facility are as follows: L_A = 0.90 m, L_B = 6.25 m, L_C = 4.6 m, and L_D = 4.2 m, so creating a quiet zone of l_B = 2.2 m, l_C = 1.6 m, and φ = 130°. The structure is lined with 28% carbon-loaded urethane

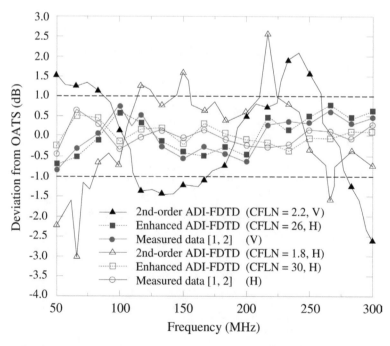

FIGURE 5.11: Deviation from OATS concerning different ADI-FDTD simulations compared to measured data for vertical (V) and horizontal (H) polarization in a 3-m partially curved semianechoic chamber.

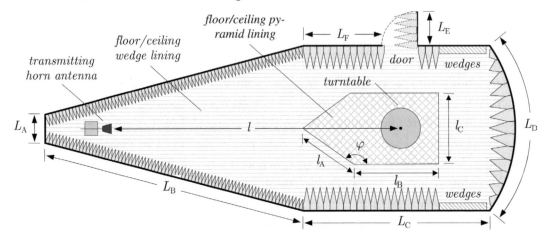

FIGURE 5.12: Top view of a tapered anechoic facility equipped with simple and curved Chebyshev pyramidal absorber arrays.

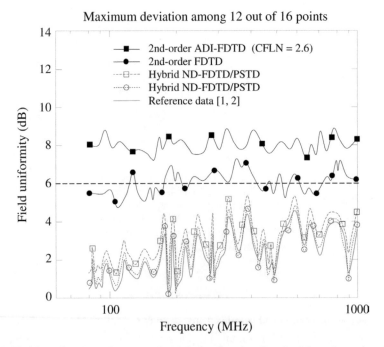

FIGURE 5.13: Field uniformity of a tapered anechoic chamber lined with various absorber media.

pyramids or arrays of alternating wedges backed by ferrite tiles and lossy ($\sigma = 185$ S/m) layers. The frequency dependence of $\varepsilon_r(\omega)$ is modeled by $\gamma_1 = 1235.07$, $\tau_1 = 51.79$ ns, $\gamma_2 = 924.736$, $\tau_2 = 75.39$ ns, and $\gamma_3 = 1081.434$, $\tau_3 = 89.12$ ns, with an analogous treatment for $\mu_r(\omega)$. Moreover, the size of the pyramids behind the quiet zone follows a curved Chebyshev regime, whereas the absorbers in the upper and lower rectangular walls have $d_A = 0.35$ m, $d_C + d = 0.95 + 0.27 = 1.22$ m. In the remaining regions, pyramids are 1/4 to 1/2 the height of the previous absorbers whereas wedges have $d_A = 0.58$ m, $d_C + d = 0.25 + 0.07 = 0.32$ m, $\alpha_1 = 25°$, and $\alpha_2 = 18°$. The transmitting device is a horn antenna with a diagonal of 0.3 m, located 5 m away from the turntable, where an elliptical EUT is considered. This challenging chamber is simulated by means of a ND-FDTD/PSTD method to consider the periodicity of the absorbers, which concludes to a $94 \times 78 \times 62$ grid with $\Delta x = 0.115$ m, $\Delta y = 0.053$ m, and $\Delta z = 0.087$ m that yield a resolution of $\lambda/8$. Figure 5.13 illustrates field uniformity and Figure 5.14 examines electric-field temporal variation. Again, the hybrid method overwhelms the second-order one (92% denser lattice with $220 \times 184 \times 168$ cells), providing that the specific facility is acceptable for immunity measurements.

Another type of EMC test facility with an improved low-frequency performance is the double-horn semianechoic chamber, described in Figure 5.15. This design uses a reduced (about 26%)

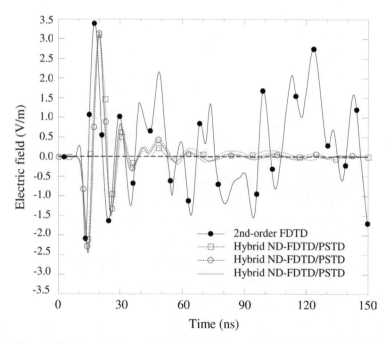

FIGURE 5.14: Electric field variation versus time inside a tapered anechoic chamber for several hybrid time-domain simulations.

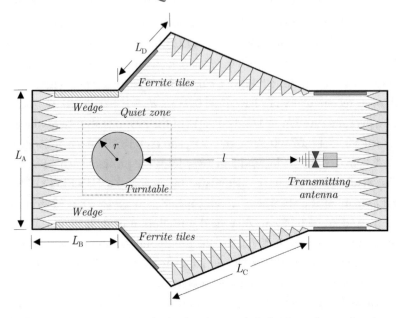

FIGURE 5.15: A double-horn semianechoic chamber with hybrid urethane absorbers and arrays of alternating wedges.

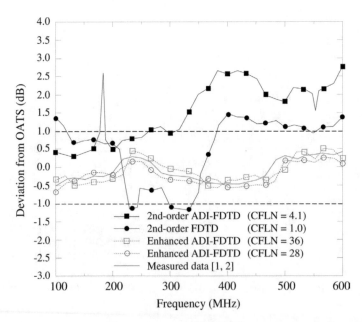

FIGURE 5.16: Deviation from OATS for diverse ADI-FDTD cases compared to measured data in a double-horn semianechoic facility (vertical polarization).

chamber volume that enforces waves to impinge on absorbers in a nearly normal direction, hence obtaining higher reflection losses. The structure has a height of 7.2 m and $L_A = 6.8$ m, $L_B = 4.1$ m, $L_C = 7.5$ m, $L_D = 3.6$ m, $l = 7$ m, whereas its 29% carbon-loaded lining combines pyramidal absorbers ($d_A = 0.35$ m, $d_C + d = 0.90 + 0.32 = 1.22$ m) with alternating wedge arrays ($d_A = 0.44$ m, $d_C + d = 0.22 + 0.06 = 0.28$ m).

Implementing the frequency-dependent treatment of (5.5) and selecting $M = 2$, $L = 2$ for the nonstandard formulae, a $94 \times 76 \times 58$ mesh is generated with $\Delta t_{max}^{ND} = 0.179$ ns and a resolution of $\lambda/10$. To characterize the chamber, deviation from OATS is depicted in Figure 5.16 for vertical polarization. The agreement of the dispersionless ADI-FDTD algorithm (CPU time, 2500 time-steps) with the measured data is excellent, unlike the outcomes of the FDTD approach. Similar deductions may be drawn from the NSA indicator of Figure 5.17 (horizontal polarization), where usual schemes use an 86% denser grid with $\Delta t_{max}^{FDTD} = 0.047$ ns and 58,000 time-steps. Similarly, Figure 5.18 confirms these conclusions through the absorber lining reflectivity at low frequencies. Enhanced accuracy is acquired as observed from the inlet figure.

The last example explores the performance of a specialized curvilinear facility, designated as hardware-in-the-loop anechoic chamber. With the shape of a pie in the azimuth plane (Figure 5.19),

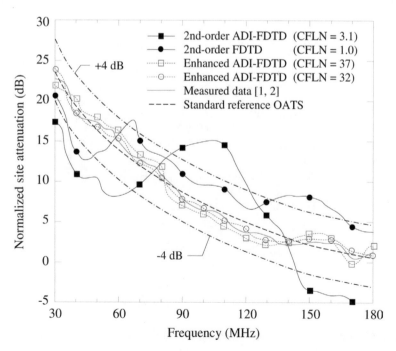

FIGURE 5.17: NSA for a double-horn semianechoic chamber (horizontal polarization).

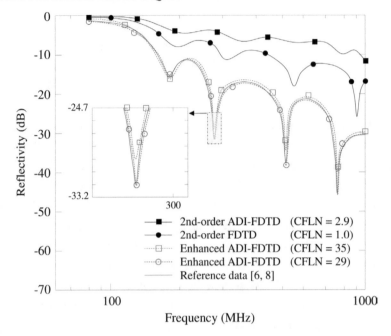

FIGURE 5.18: Reflectivity of a combined urethane pyramid/alternating wedge lining via various ADI-FDTD layouts.

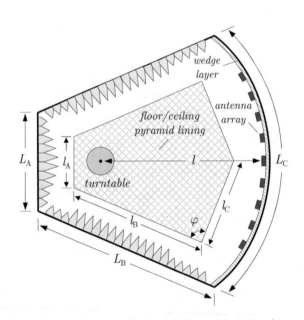

FIGURE 5.19: A hardware-in-the-loop anechoic chamber with twisted pyramidal absorbers and a wedge layer.

this structure illuminates the EUT through an array of antennas mounted at certain positions on the circularly shaped wall. Its basic use focuses on immunity measurements concerning multiple objects of different size. Therefore, it has relatively large dimensions, i.e., $L_A = 3.8$ m, $L_B = 10.2$ m, $l = 6.5$ m, and a height of 8.6 m. The chamber is equipped with 35° twisted pyramids ($d_A = 0.61$ m, $d_C + d = 2.08 + 0.36 = 2.44$ m, 28% carbon-loaded), backed by 8.4-mm ferrite tiles and lossy ($\sigma = 2.74$ S/m) layers and a set of thin wedges. Relative dielectric permittivity, $\varepsilon_r(\omega)$ is approximated by $\gamma_1 = 458.12$, $\tau_1 = 76.43$ ns, $\gamma_2 = 671.958$, $\tau_2 = 44.72$ ns and $\gamma_3 = 816.559$, $\tau_3 = 97.57$ ns fitting coefficients, whereas the hybrid ADI-FDTD/FETD method is selected for our simulations. The resulting domain comprises $86 \times 64 \times 52$ cells with $\Delta x = 0.118$ m, $\Delta y = 0.059$ m, $\Delta z = 0.165$ m, and 86,720 finite elements. Field uniformity, given in Figure 5.20, demonstrates that the accuracy of the technique is indeed very sufficient despite the curved wall and the increased number of reflected waves. Toward this direction, Figure 5.21 presents the temporal variation of electric field. A worth-mentioning issue for the common ADI-FDTD and FDTD methods is the large number of artificial reflections aroused after the appearance of the main pulse, although both approaches use a $192 \times 172 \times 158$ lattice with $\Delta t_{\max}^{\text{FDTD}} = 0.041$ ns and small CFLN. These defects—practically alleviated by the

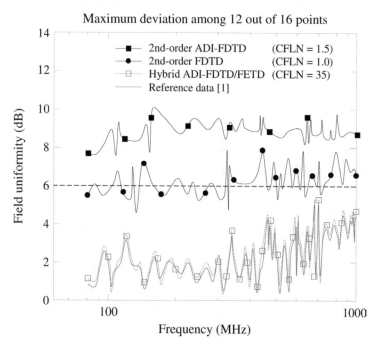

FIGURE 5.20: Field uniformity of a hardware-in-the-loop anechoic chamber lined with 35° twisted pyramidal absorbers and wedges.

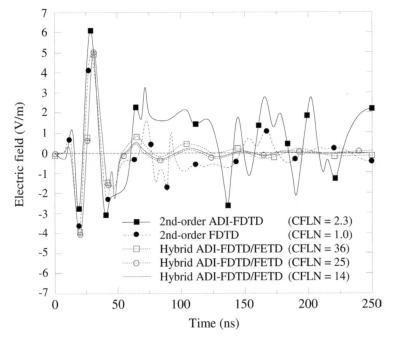

FIGURE 5.21: Temporal variation of the electric field inside a hardware-in-the-loop anechoic facility.

chosen formulation—are attributed to the inherent dispersion mechanisms and the abrupt scattering interactions near the EUT.

5.4.3   Efficiency Considerations

Having confirmed the advantages of the different time-domain algorithms—particularly in hybrid variants—as a rigorous tool for broadband chamber performance evaluations, numerical analysis will now deal with the significant issues of design and efficiency. The basic objective is the attainment of the facility's suitability for measurements through the optimization of certain features. These are the type, the constitutive parameters and the size of the lining, the number and distance of the absorbers from the EUT as well as the dimensions and shape of the walls. Even though any change in the geometrical aspects of an existing chamber is rather difficult and high-priced, this issue will be investigated as a means of determining the optimal configuration for new test facilities. Because the required grid resolutions along with the total amount of time-steps can now be radically reduced, it is obvious that such simulations are fairly feasible.

The most convenient way to enhance the performance of already built chambers is the careful adjustment of their absorption capabilities. Recall the structure of Figure 5.2c with $L_A = 7.6$ m, $2L_B =$

2.2 m, $2L_C = 4.8$ m, and a height of 5.6 m, where the five absorbers of Table 5.3 are studied via the enhanced ADI-FDTD method.

The discretization procedure results in an $82 \times 68 \times 62$ mesh with $\Delta t_{max}^{ND} = 0.168$ ns. Note that all linings are backed by thin ferrite tiles and a dielectric layer whose contribution to the facility's low-frequency operation is, indeed, crucial. The transmitting/receiving antenna distance is set to 3 m, while a spherical EUT of radius 0.35 m is also involved. To guarantee fast and precise predictions, the curvilinear regions are modeled with $M = 3$, $L = 3$, and CFLN = 31. Figure 5.22a shows the reflectivity of these media for several conductivities. Apparently, the hybrid urethane absorber E yields the best results, because it annihilates the tangential field components near the shielding walls, mainly from 30 to 150 MHz. Simple pyramidal absorber B, on the other hand, exhibits a rather low reflectivity level, although its σ value is quite high. Furthermore, the NSA of the chamber is given in Figure 5.22b, where, again, absorber E is proven to be the most effective with the 45° twisted pyramids (absorber D) showing an equivalent performance and absorbers A and B lacking to fulfill the ±4 dB limit. Analogous results are summarized in Table 5.4, which includes several ADI-FDTD aspects and the maximum deviation from OATS at the frequency of 350 MHz. It is important to indicate the satisfaction of the ±1 dB criterion, the markedly coarse grids despite the complex test facility components, and the shorter simulations achieved by the improved unconditionally stable technique.

The impact of a consistent optimization procedure is next explored via the structures of Figures 5.6 and 5.12. Let us initially examine the former chamber with a width of 6.8 m, $L_A = 7.4$ m, $L_B = 5.6$ m, $L_C = 6.5$ m, $L_D = 2.8$ m, $L_E = 9.2$ m, and $L_F = 12.5$ m. The diameters of its two reflectors are 2.5 and 3.8 m, respectively, and the optimized absorber arrays comprise 40° twisted pyramids

TABLE 5.3: Structural and material properties of different absorber array linings

ABSORBER ARRAY (CARBON LOADING)	DIMENSIONS (m)	FERRITE TILE GAP (mm)
A Alternating wedges (27%)	$d_A = 0.47$, $d_C + d = 0.26 + 0.08 = 0.34$	0.28
B Simple urethane pyramids (28%)	$d_A = 0.52$, $d_C + d = 1.96 + 0.48 = 2.44$	0.43
C 35° twisted pyramids (30%)	$d_A = 0.36$, $d_C + d = 0.96 + 0.26 = 1.22$	0.53
D 45° twisted pyramids (29%)	$d_A = 0.52$, $d_C + d = 1.47 + 0.36 = 1.83$	0.56
E Hybrid urethane pyramids (26%)	$d_A = 0.61$, $d_C + d = 2.14 + 0.30 = 2.44$	0.22

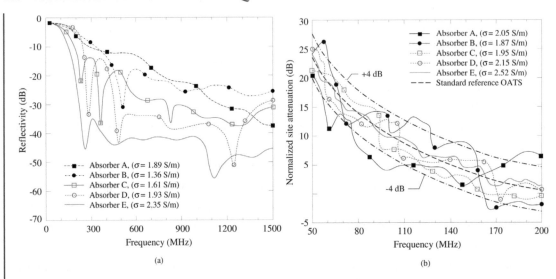

FIGURE 5.22: Performance optimization of a 3-m partially curved semianechoic chamber lined with the absorbers of Table 5.3 through (a) media reflectivity and (b) NSA.

($d_A = 0.56$ m, $d_C + d = 2.08 + 0.36 = 2.44$ m, 30% carbon-loaded). To accomplish wideband realizations, the lining's relative permittivity is approximated by $\gamma_1 = 563.84$, $\tau_1 = 42.51$ ns, $\gamma_2 = 718.463$, $\tau_2 = 74.92$ ns, and $\gamma_3 = 1034.124$, $\tau_3 = 117.35$ ns, whereas for the numerical analysis the TLM technique is used. The first item of the study is the influence of the gap between the 5.2-mm-thick ferrite tiles used for the backing of the absorbers. Electric field variation is calculated in a rectangular region, 1.2 m above the turntable, for a 1.45-mm gap (Figure 5.23a) and a 0.25-mm gap (Figure 5.23b) at 200 MHz. Although the difference of the two gaps is very small, its effect is exceptionally serious, because a properly tuned choice can lead to an unperturbed quiet zone without spurious wave patterns. Indeed, as observed, the environment of Figure 5.23a is completely unreliable for any type of EMC test in contrast to that of Figure 5.23b. Next, for the chamber of Figure 5.12 ($L_A = 0.85$ m, $L_B = 6.5$ m, $L_C = 5.2$ m, $L_D = 4.6$ m), several simple pyramidal absorbers at a Chebyshev layout are used. The time-domain algorithm studies both optimized and nonoptimized (incorrect carbon loading and $N \leq 2$ frequency-dependent approximation) versions. Figure 5.24 illustrates field uniformity from 300 MHz to 5 GHz, where the former absorbers produce a promising estimation of the chamber's measurement suitability, unlike the latter, which exceed the 6-dB threshold.

Finally, the issue of optimal geometric characteristics is examined for the construction of new chambers with a prefixed quiet zone. Therefore, assume the facility of Figure 5.12, which involves an systematically designed combination of hybrid urethane absorbers ($d_A = 0.57$ m, $d_C + d = 1.35 + 0.48 = 1.83$ m) and arrays of alternating wedges ($d_A = 0.38$ m, $d_C + d = 0.28 + 0.08$

TABLE 5.4: Comparison of ADI-FDTD implementation aspects and maximum deviation from OATS at 350 MHz for a 3-m partially curved semianechoic chamber lined with the absorbers of Table 5.3

METHOD	ABSORBER	CFLN	LATTICE	REDUCTION (%)	MEMORY (MB)	TIME-STEPS	CPU TIME (h)	MAXIMUM DEVIATION (dB)
Second-order	A	1.37	$170 \times 162 \times 204$		427	68,522	59.2	4.561
Enhanced		28	$82 \times 68 \times 84$	91.67	44	2468	4.6	$0.674 < +1$
Second-order	B	1.46	$164 \times 156 \times 192$		415	59,784	49.8	-3.812
Enhanced		35	$76 \times 64 \times 80$	92.08	35	2371	3.7	$-0.915 > -1$
Second-order	C	2.23	$182 \times 170 \times 200$		441	72,364	64.3	2.179
Enhanced		26	$88 \times 76 \times 92$	90.14	50	2892	3.2	$0.503 < +1$
Second-order	D	2.68	$174 \times 160 \times 198$		422	62,938	55.1	5.124
Enhanced		38	$86 \times 74 \times 92$	91.32	37	2180	4.4	$0.414 < +1$
Second-order	E	3.19	$188 \times 176 \times 212$		456	82,596	70.5	-4.987
Enhanced		41	$88 \times 78 \times 94$	90.25	58	2046	3.8	$-0.856 > -1$

Electric field (V/m)

Electric field (V/m)

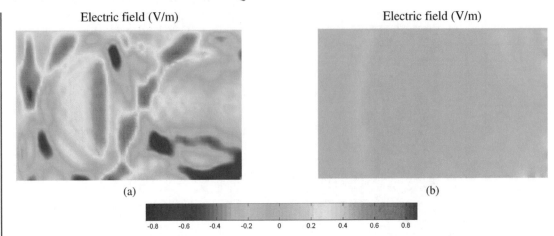

(a)

(b)

-0.8   -0.6   -0.4   -0.2   0   0.2   0.4   0.6   0.8

FIGURE 5.23: Electric field amplitude at 200 MHz through a rectangular area, 1.2 m above the turn-table, in a dual reflector anechoic facility with 40° twisted pyramids backed by 5.2-mm-thick ferrite tiles with (a) a 1.45-mm and (b) a 0.25-mm gap.

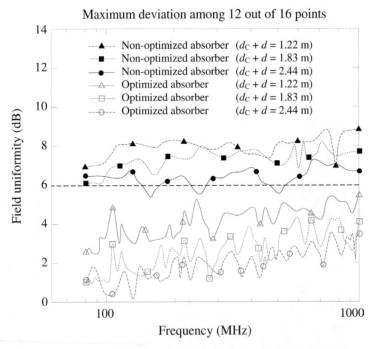

FIGURE 5.24: Field uniformity of a tapered anechoic facility lined with various optimized and nonoptimized simple pyramids.

= 0.36 m). Selecting the values for the height of the structure and for two of its dimensions, the variation of the remaining ones is explored in the form of ratios. The distance between the transmitting and receiving antenna is an additional degree of freedom, whereas the domain receives the most accurate discretizations by means of the ND-FDTD/FETD method. Under these assessments and a height of 5.2 m, Figure 5.25 presents electric-field temporal evolution for $L_D = 4.8$ m and $L_C = 5.4$ m. Notice that slight changes in the L_D/L_A and L_C/L_B ratios generate different models with contradicting properties. More specifically, it seems that the chamber's suitability for immunity measurements increases if L_D/L_A receives larger values and L_C/L_B smaller values. This rule of thumb is also validated by the facility of Figure 5.15, where the two straight walls forming the double horn on each side of the structure have been replaced with curved parts. Using the same absorbers as above, a height of 6.8 m and $L_A = 6.4$ m, $L_B = 4.8$ m, the hybrid method examines various ratios. Results for the deviation from OATS are demonstrated in Figure 5.26. It is evident that the performance of the curvilinear chamber is sensitive to certain geometric choices, which can produce some very satisfactory designs.

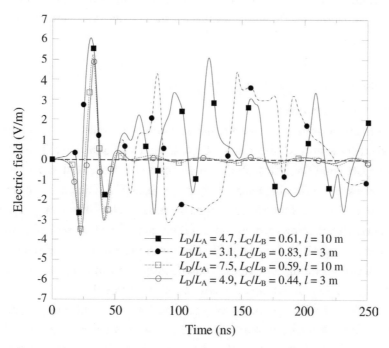

FIGURE 5.25: Efficiency optimization of a tapered anechoic chamber for different dimension ratios and transmitting/receiving antenna distance via electric-field temporal variation.

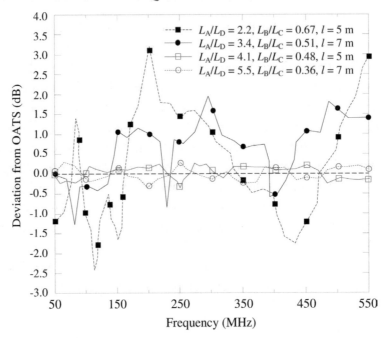

FIGURE 5.26: Efficiency optimization of a curvilinear double horn semianechoic chamber for different dimension ratios and transmitting/receiving antenna distance via deviation from OATS.

5.5   REVERBERATION CHAMBERS

5.5.1   Basic Configurations

A reverberation chamber consists of a rectangular enclosure with walls and a mode-stir paddle (also known as stirrer) that disrupts RF fields inside the test facility (Figure 5.27a) [17–31, 58]. Actually, it emulates free-space conditions through the process of mode stirring and therefore it can be used as an alternate means for performing radiated immunity tests. The fundamental operation of such a chamber lies on the existence of multimode resonance mixing. Notwithstanding its design simplicity, the facility entails a rather involved procedure for its study. This is accredited to the requirement for sound theoretical analysis to describe field behavior in the enclosure and the need to correlate test results with actual operating conditions. Consequently, it becomes apparent that time-domain numerical methods can be proven a very powerful assistant to this aim.

A reverberation chamber produces an environment where the radiated field is uniform throughout the volume of the enclosure walls and hence it provides random, complex, real-world conditions for immunity and emission testing. Moreover, it can be used for radiated susceptibility measurements with certain advantages over other structures. This is due to the excellent isolation

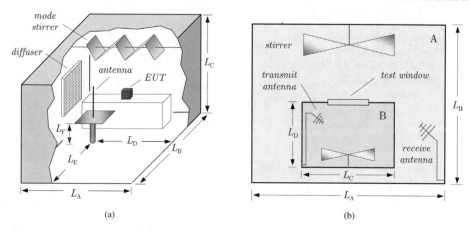

FIGURE 5.27: (a) A general mode-stirred reverberation facility. (b) A nested reverberation chamber environment.

from the external electromagnetic environment it guarantees. Nonetheless, one disadvantage is the link of the obtained measurements to realistic situations, because the polarization properties of the EUT are unknown. To circumvent this shortcoming, a set of supplementary metallic vanes are inserted in the chamber from adjacent walls and rotated at different speeds around an axis perpendicular to the wall. The temporal variation of the chamber's geometry along with the rotation of the vanes produces a constant mixing of modes with the same statistical distribution. Now, regardless of where the EUT is located, field intensity is kept uniform.

Bearing in mind the preceding remarks, reverberation chambers—implemented also in the nested environment [27] of Figure 5.27b—are essentially used for automotive applications, shielding effectiveness validation, certification of international standards, commercial avionics, and large-EUT hardness testing at high field strengths. Toward this direction, the desired field intensity for immunity tests is launched with a signal generator, attenuator, and amplifier, while the stirrer vanes are kept in continual rotation.

5.5.2   Numerical Investigation

Proceeding to the numerical characterization of reverberation chambers, the enhanced ADI-FDTD formulation and an optimized TLM technique evaluate the overall field distribution and the shielding effectiveness of various media. Both single and nested environments are studied. Their performance is further advanced by quadratic residue diffusers, namely, a set of periodically built phase gratings that diffract incident waves in auxiliary directions. Stirrer movements are discretely modeled with an interval of 25°, whereas for each of its positions we sample 50 time instants. The dimensions

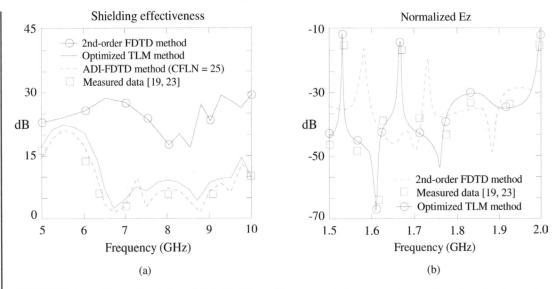

FIGURE 5.28: Normalized E_z and (b) shielding effectiveness of a composite medium in a single mode-stirred reverberation chamber.

of our first application, shown in Figure 5.27a, are $L_A = 3.55$ m, $L_B = 4.86$ m, $L_C = 3.72$ m, $L_D = 1.7$ m, $L_E = 1.8$ m, and $L_F = 0.5$ m. The computational domain is divided in $64 \times 80 \times 68$ cells, thus attaining the coarse, yet ample, resolution of $\lambda/6$. Figure 5.28a presents the shielding effectiveness of a carbon fiber with a thickness of 1 mm, whereas Figure 5.28b illustrates the normalized E_z field

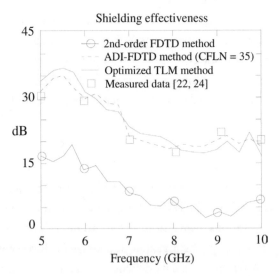

FIGURE 5.29: Shielding effectiveness for a composite material in a nested reverberation chamber setup.

component for a CFLN = 25, as well. In fact, the improved ADI-FDTD algorithm overcomes the second-order FDTD one, whose deviation from the measurements is rather significant.

Subsequently, the nested configuration of Figure 5.27b is validated. The larger chamber has a length of 5.2 m, L_A = 3.86 m, L_B = 4.10 m, and the smaller one a length of 2.64 m, L_C = 1.52 m, L_D = 1.96 m with an aperture size of 0.35 × 0.35 m. Figure 5.29 displays the shielding effectiveness of a 1.5-mm-thick fiber-glass-fiber material for a 86 × 78 × 92 mesh; 90% coarser than its FDTD analogue. Again, the ADI-FDTD and optimized TLM algorithms achieve promising broadband simulations without crucial dispersion errors.

5.6   TEM AND GTEM CELLS

There exist two basic types of cells: the transverse electromagnetic (TEM) and the gigahertz transverse electromagnetic (GTEM) cell [32–36]. Their purpose is to verify the EMC qualifications of products that are physically small (up to one-third the volume of the cell) and especially of certain parts or components. The TEM (or Crawford) cell is a small enclosure used in normal laboratory environments for both emission analysis and radiated immunity. In its typical version, it is a rectangular coaxial transmission *line* resembling a stripline with outer conductors closed and joined together, as shown Figure 5.30a. Both ends of the rectangular section are tapered to a 50-Ω coaxial cable, while the central conductor and the outer shield (top and bottom plates plus sides) facilitate the propagation of electromagnetic energy from one end of the cell to the other in the TEM mode. To conduct a measurement, the EUT—electrically isolated via a dielectric spacer—is placed on the rectangular portion of the transmission line between the top or bottom plate and the central conductor. It is stressed that the presence of the closed outer shell provides an effective shield to obstruct electromagnetic waves from both entering and leaving the cell. Indisputably, an advantage of a TEM cell is its small size, large flexibility, low cost, and lack of need for a high-power amplifier. Also, no additional shielding is required to attenuate external radiated fields. Nevertheless, owing to its particular geometry and confined dimensions, the TEM cell can not adequately support high operating frequencies or extended spectra. Moreover, if the EUT size is not small, the RF field inside the cell is essentially shorted out at certain locations, causing a higher intensity of field strengths to occur. As an example of numerical modeling in the time domain, Figure 5.30b gives the shielding effectiveness of a 2-mm-thick low-frequency carbon material, measured in a 1.5 × 1.5 m TEM cell and simulated by means of the FETD method. Evidently, the results are in very good agreement with the measured data, so proving the efficiency of finite elements to handle such EMC problems.

Similar in its function, but with certain benefits over a regular TEM facility, is the GTEM cell depicted in Figure 5.31a. The most important one is that the restriction on the upper frequency limit is eliminated by tapering the transmission line continuously outward from the feed point to a termination system. Actually, the tapered point and the absorber lining at the larger end of the cell

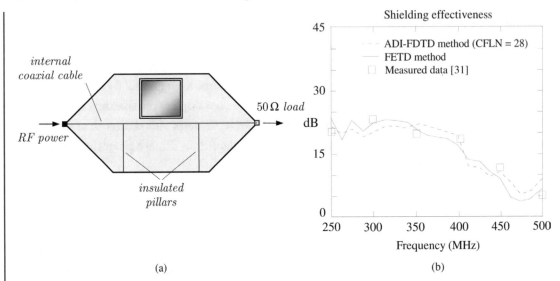

(a)   (b)

FIGURE 5.30: (a) Geometry of a TEM cell. (b) Shielding efficiency of a low-frequency carbon material.

are the two factors that allow this enclosure to operate well into the gigahertz range without the need of a shielded facility. An indicative illustration of a numerical time-domain study is provided in Figure 5.31b, where the shielding effectiveness of a 1.4-mm-thick fiber-carbon-fiber material is computed through the TLM method. The dimensions of the GTEM cell are $L_A = 2.5$ m and $L_B = 1.2$ m, and its absorbers are 28% carbon-loaded urethane pyramids with $d_A = 0.47$ m and $d_C + d =$

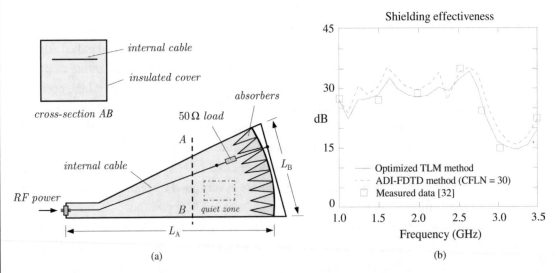

(a)   (b)

FIGURE 5.31: (a) Geometry of a typical GTEM cell. (b) Shielding efficiency of a fiber-carbon-fiber material.

0.26 + 0.08 = 0.34 m. From the outcomes, one can easily realize that the selected algorithm copes very successfully with the entire structure, without inducing any nonphysical parasites.

REFERENCES

1. L. H. Hemming, *Electromagnetic Anechoic Chambers: A Fundamental Design and Specification Guide*. New York: IEEE Press and Wiley Interscience, 2002.

2. M. I. Montrose and E. D. Nakauchi, *Testing for EMC Compliance—Approaches and Techniques*. New York: IEEE Press and Wiley Interscience, 2004.

3. R. Janaswamy, "Oblique scattering from lossy periodic surfaces with application to anechoic chamber absorbers," *IEEE Trans. Antennas Propagat.*, vol. 40, no. 2, pp. 162–169, Feb. 1992. doi:10.1109/8.127400

4. C. L. Holloway and E. F. Kuester, "Modeling semi-anechoic electromagnetic measurement chambers," *IEEE Trans. Electromagn. Compat.*, vol. 38, no. 1, pp. 79–84, Feb. 1996. doi:10.1109/15.485700

5. A. Orlandi, "Multipath effects in semi-anechoic at low frequencies: A simplified prediction model based on image theory," *IEEE Trans. Electromagn. Compat.*, vol. 38, no. 3, pp. 478–483, Aug. 1996. doi:10.1109/15.536078

6. C. L. Holloway, R. R. DeLyser, R. F. German, P. M. McKenna, and M. Kanda, "Comparison of electromagnetic absorber used in anechoic and semi-anechoic chambers for emissions and immunity testing of digital devices," *IEEE Trans. Electromagn. Compat.*, vol. 39, no. 1, pp. 33–47, Feb. 1997. doi:10.1109/15.554693

7. C. Bornkessel and W. Wiesbeck, "Numerical analysis and optimization of anechoic chambers for EMC testing," *IEEE Trans. Electromagn. Compat.*, vol. 38, no. 3, pp. 499–506, Aug. 1996. doi:10.1109/15.536082

8. B. Fourestié, Z. Altman, and M. Kanda, "Efficient detection of resonances in anechoic chambers using the matrix pencil method," *IEEE Trans. Electromagn. Compat.*, vol. 42, no. 1, pp. 1–5, Feb. 2000. doi:10.1109/15.831699

9. A. R. Ruddle and D. D. Ward, "Numerical modeling as a tool for cost effective design and optimization of EMC test chambers," in *Proc. 4th European Symp. Electromagn. Compat.*, Brugge, Belgium, Sept. 2000, pp. 137–142.

10. B. Archambeault, P. Valentino, and E. Schumann, "Evaluating semi-anechoic rooms using an accurate and ultra-repeatable source," in *Proc. IEEE Int. Symp. Electromagn. Compat.*, Montreal, Canada, Aug. 2001, pp. 958–963. doi:10.1109/ISEMC.2001.950520

11. C. L. Holloway, P. M. McKenna, R. A. Dalke, R. A. Perala, and C. L. Devor, "Time-domain modeling, characterization, and measurements of anechoic and semi-anechoic electromagnetic

test chambers," *IEEE Trans. Electromagn. Compat.*, vol. 44, no. 1, pp. 102–118, Feb. 2002. doi:10.1109/15.990716

12. L. Musso, F. Canavero, B. Demoulin, and V. Berat, "Radiated immunity testing of a device with an external wire: Repeatability of reverberation chamber results and correlation with anechoic chamber results," in *Proc. IEEE Symp. Electromagn. Compat.*, Boston, USA, Aug. 2003, pp 828–833. doi:10.1109/ISEMC.2003.1236715

13. N. V. Kantartzis and T. D. Tsiboukis, "Unconditionally stable numerical modeling and broadband optimization of arbitrarily-shaped anechoic and reverberating chambers," in *Proc. EMC Europe 2004 Int. Symp. Electromagn. Compat.*, Eindhoven, The Netherlands, Sept. 2004, pp. 48–53.

14. N. V. Kantartzis, T. T. Zygiridis, and T. D. Tsiboukis, "An unconditionally stable higher-order ADI-FDTD technique for the dispersionless analysis of generalized 3-D EMC structures," *IEEE Trans. Magn.*, vol. 40, no. 2, pp. 1436–1439, Mar. 2004. doi:10.1109/TMAG.2004.825289

15. N. V. Kantartzis and T. D. Tsiboukis, "Wideband numerical and performance optimisation of arbitrarily-shaped anechoic chambers via an unconditionally stable time-domain technique," *Electrical Eng.*, vol. 81, pp. 55–81, 2005. doi:10.1007/s00202-004-0252-4

16. N. V. Kantartzis, T. D. Tsiboukis, and E. E. Kriezis, "An explicit weighted essentially non-oscillatory time-domain algorithm for 3-D EMC applications with arbitrary media," *IEEE Trans. Magn.*, vol. 42, no. 4, pp. 803–806, Apr. 2006.

17. D. A. Hill, "Electronic mode stirring for reverberation chambers," *IEEE Trans. Electromagn. Compat.*, vol. 36, no. 4, pp. 294–299, Nov. 1994. doi:10.1109/15.328858

18. D. A. Hill, "Plane wave integral representation for fields in reverberation chambers," *IEEE Trans. Electromagn. Compat.*, vol. 40, no. 3, pp. 209–217, Aug. 1998. doi:10.1109/15.709418

19. C. F. Bunting, K. J. Moeller, C. J. Reddy, and S. A. Scearce, "A two-dimensional finite-element analysis of reverberation chambers," *IEEE Trans. Electromagn. Compat.*, vol. 41, no. 4, pp. 280–289, Nov. 1999. doi:10.1109/15.809794

20. N. K. Kouveliotis, P. T. Trakadas, and C. N. Capsalis, "Examination of field uniformity in vibrating intrinsic reverberation chamber using the FDTD method," *IEE Electron. Lett.*, vol. 38, no. 3, pp. 109–110, 2002. doi:10.1049/el:20020076

21. S. Silfverskiöld, M. Bäckström, J. Lorén, "Microwave field-to-wire coupling measurements in anechoic and reverberation chambers," *IEEE Trans. Electromagn. Compat.*, vol. 44, no. 1, pp. 222–232, Feb. 2002. doi:10.1109/15.990729

22. U. Carlberg, P.-S. Kildal, A. Wolfgang, O. Sotoudeh, and C. Orlenius, "Calculated and measured absorption cross sections of lossy objects in reverberation chambers," *IEEE Trans. Electromagn. Compat.*, vol. 46, no. 2, pp. 146–154, May 2004. doi:10.1109/TEMC.2004.826878

23. P. Corona, G. Ferrara, and M. Migliaccio, "Generalized stochastic field model for reverberating chambers," *IEEE Trans. Electromagn. Compat.*, vol. 46, no. 4, pp. 655–660, Nov. 2004. doi:10.1109/TEMC.2004.837831

24. C. Bruns and R. Vahldieck, "A closer look at reverberation chambers—3-D simulation and experimental verification," *IEEE Trans. Electromagn. Compat.*, vol. 47, no. 3, pp. 612–626, Aug. 2005. doi:10.1109/TEMC.2005.850677

25. L. R. Arnaut, "On the maximum rate of fluctuation in mode-stirred reverberation," *IEEE Trans. Electromagn. Compat.*, vol. 47, no. 4, pp. 781–804, Nov. 2005. doi:10.1109/TEMC.2005.859061

26. J. Clegg, A. C. Marvin, J. F. Dawson, and S. J. Porter, "Optimization of stirrer designs in a reverberation chamber," *IEEE Trans. Electromagn. Compat.*, vol. 47, no. 4, pp. 824–832, Nov. 2005. doi:10.1109/TEMC.2005.860561

27. C. L. Holloway, D. A. Hill, J. M. Ladbury, and G. Koepke, "Requirements for an effective reverberation chamber: Unloaded or loaded," *IEEE Trans. Electromagn. Compat.*, vol. 48, no. 1, pp. 187–194, Feb. 2006. doi:10.1109/TEMC.2006.870709

28. G. Gradoni, F. Moglie, A. P. Pastore, and V. M. Primiani, "Numerical and experimental analysis of the field to enclosure coupling in reverberation chamber and comparison to anechoic chamber," *IEEE Trans. Electromagn. Compat.*, vol. 48, no. 1, pp. 203–211, Feb. 2006. doi:10.1109/TEMC.2006.870805

29. N. Wellander, O. Lundén, and M. Bäckström, "Experimental investigation and mathematical modeling of design parameters for efficient stirrers in mode-stirred reverberation chambers," *IEEE Trans. Electromagn. Compat.*, vol. 49, no. 1, pp. 94–103, Feb. 2007.

30. A. Coates, H. G. Sasse, D. E. Coleby, A. P. Duffy, and A. Orlandi, "Validation of a three-dimensional transmission line matrix (TLM) model implementation of mode-stirred reverberation chamber," *IEEE Trans. Electromagn. Compat.*, vol. 49, no. 4, pp. 734–744, Nov. 2007.

31. G. Orjubin, E. Richalot, O. Picon, and O. Legrand, "Chaoticity of a reverberation chamber assessed from the analysis of modal distributions obtained by FEM," *IEEE Trans. Electromagn. Compat.*, vol. 49, no. 4, pp. 762–771, Nov. 2007.

32. J. P. Karst, C. Groh, and H. Garbe, "Calculable field generation using TEM cells applied to the calibration of a novel E-field probe," *IEEE Trans. Electromagn. Compat.*, vol. 44, no. 1, pp. 59–71, Feb. 2002. doi:10.1109/15.990711

33. T.-H. Loh and M. J. Alexander, "A method to minimize emission measurement uncertainty of electrically large EUTs in GTEM cells and FARs above 1 GHz," *IEEE Trans. Electromagn. Compat.*, vol. 48, no. 4, pp. 634–640, Nov. 2006.

34. X. T. I. Ngu, A. Nothofer, D. W. P. Thomas, and C. Christopoulos, "A complete model for simulating magnitude and phase if emissions from a DUT placed inside a GTEM cell," *IEEE Trans. Electromagn. Compat.*, vol. 49, no. 2, pp. 285–293, May 2007.

35. P. E. Fornberg and C. L. Holloway, "A comparison of the currents induced on an EUT in a TEM cell to those induced in a free-space environment," *IEEE Trans. Electromagn. Compat.*, vol. 49, no. 3, pp. 474–484, Aug. 2007.

36. S. Deng, T. Hubing, and D. G. Beetner, "Characterizing the electric field coupling from IC heatsink structures to external cables using TEM cell measurements," *IEEE Trans. Electromagn. Compat.*, vol. 49, no. 4, pp. 785–791, Nov. 2007.

37. T. Namiki, "A new FDTD algorithm based on ADI method," *IEEE Trans. Microwave Theory Tech.*, vol. 47, no. 10, pp. 2003–2007, Oct. 1999. doi:10.1109/22.795075

38. F. Zheng, Z. Chen, and J. Zhang, "A finite-difference time-domain method without the Courant stability conditions," *IEEE Microwave Guided Wave Lett.*, vol. 9, no. 11, pp. 441–443, Nov. 1999. doi:10.1109/75.808026

39. E. Hu and W. J. R. Hoefer, "Performance of three-dimensional graded ADI-FDTD algorithm," in *Proc. IEEE Int. Microwave Symp.*, Phoenix, USA, May 2001, pp 901–904. doi:10.1109/MWSYM.2001.967037

40. S. Garcia, T. Lee, and S. Hagness, "On the accuracy of the ADI-FDTD method," *IEEE Antennas Wireless Prop. Lett.*, vol. 1, pp. 31–34, 2002. doi:10.1109/LAWP.2002.802583

41. M. Darms, R. Schuhmann, H. Spachmann, and T. Weiland, "Dispersion and asymmetry effects of ADI-FDTD," *IEEE Microwave Wireless Components Lett.*, vol. 12, pp. 491–493, Dec. 2002.

42. A. P. Zhao, "Analysis of the numerical dispersion of the 2-D alternating-direction implicit FDTD method," *IEEE Trans. Microwave Theory Tech.*, vol. 50, pp. 1156–1164, 2002.

43. S. Staker, C. Holloway, A. Bhobe, and M. Piket-May, "ADI formulation of the FDTD method: Algorithm and material dispersion implementation," *IEEE Trans. Electromagn. Compat.*, vol. 45, pp. 156–166, May 2003. doi:10.1109/TEMC.2003.810815

44. S. G. García, R. G. Rubio, A. R. Bretones, and R. G. Martín, "Extension of the ADI-FDTD method to Debye media," *IEEE Trans. Antennas Propag.*, vol. 51, no. 11, pp. 3183–3186, Nov. 2003. doi:10.1109/TAP.2003.818770

45. M. Chai, T. Xiao, and Q. H. Liu, "Conformal method to eliminate the ADI-FDTD straircasing errors," *IEEE Trans. Electromagn. Compat.*, vol. 48, no. 2, pp. 273–281, May 2006. doi:10.1109/TEMC.2006.874084

46. N. V. Kantartzis and T. D. Tsiboukis, *Higher-Order FDTD Schemes for Waveguide and Antenna Structures*, San Rafael, CA: Morgan & Claypool Publishers, 2006. doi:10.2200/S00018ED1V01Y200604CEM003

47. R. E. Mickens, Ed., *Applications of Nonstandard Finite Difference Schemes*. Singapore: World Scientific, 2000.

48. J. B. Cole, "High-accuracy Yee algorithm based on nonstandard finite differences: New developments and verifications," *IEEE Tran,. Antennas Propag.*, vol. 50, pp. 1185–1191, 2002. doi:10.1109/TAP.2002.801268

49. S. Loredo, M. R. Pino, F. Las-Heras, and T. K. Sarkar, "Echo identification and cancellation techniques for antenna measurement in non-anechoic test sites," *IEEE Antennas Propagat. Mag.*, vol. 46, no. 1, pp. 100–107, Feb. 2004. doi:10.1109/MAP.2004.1296154

50. *Electromagnetic Compatibility (EMC)—Part 4: Testing and Measurement Techniques—Sec. 4: Radiated Radio-Frequency, Electromagnetic Field Immunity Test*. IEC Standard 61000-4-3, 1998.

51. *American National Standard Guide for the Computation of Errors in Open-Area Test Site Measurements*. American National Standards Institute, ANSI C63.6, New York: IEEE, 1996.

52. *Methods of Measurement of Radio-Noise Emissions from Low-Voltage Electrical and Electronic Equipment in the Range of 9 kHz to 40 GHz*. American National Standard Institute, ANSI C63.4, New York: IEEE, 2000.

53. *Information Technology Equipment—Radio Disturbance Characteristics—Limits and Methods of Measurement*. International Electrotechnical Commission Standard CISPR22, 1997.

54. *American National Standard for Electromagnetic Compatibility—Radiated Emission Measurements in Electromagnetic Interference (EMI) Control—Calibration of Antennas*. American National Standards Institute, ANSI C63.5, New York: IEEE, 1998.

55. *Electromagnetic Compatibility Basic Emissions Standard—Part 3: Emission Measurements in Fully Anechoic Rooms*. CENELEC Standard EN 50147-3, Brussels, 2000.

56. C. A. Balanis, *Antenna Theory: Analysis and Design*. New York, USA: John Wiley and Sons, 1997.

57. N. V. Kantartzis and C. S. Antonopoulos, "Design of hybrid time-domain schemes with optimal gridding density and material-interface sensitivity for large-scale EMC problems," *IEEE Trans. Magn.*, vol. 44, no. 6, pp. 1462-1465, June 2008.

58. *Electromagnetic Compatibility (EMC)—Part 4: Testing and Measurement Techniques—Sec. 21: Reverberation chamber test methods*. IEC Standard 61000-4-21, 1998.

CHAPTER 6

Large-Scale EMC and Electrostatic Discharge Problems

6.1 INTRODUCTION

The size of electromagnetic applications and devices is uncontradictably a crucial parameter in their computational modeling, especially via a full-wave numerical technique. Notwithstanding the impressive evolution of top-end PCs or workstations, both in clock speed and storage memory, there are still several realistic problems whose solution remains unaffordable. Actually, large-scale structures are an indispensable part of the EMC realm and thus any attempt for their reasonable numerical manipulation has to be carefully established. Having, so far, presented and validated most of the well-known time-domain methodologies, the theoretical analysis of this chapter is devoted to the description of a hybrid FDTD/PSTD algorithm with notable gridding capabilities. Further enhancement is accomplished via a domain decomposition concept that produces easy-to-model modular subregions of the main space. In addition, special attention is paid to the topic of electrostatic discharge (ESD) that is considered among the basic origins of undesired disturbances and malfunctions. Because this natural phenomenon occurs suddenly and cannot be easily predicted, its consequences may be very serious, if not detrimental, for many systems. Therefore, its quantification requires a cautious investigation.

6.2 HYBRID TIME-DOMAIN SCHEMES WITH OPTIMAL GRIDDING DENSITY

A critical aspect in the simulation of large-scale EMC problems is the presence of fine, as regards the minimum wavelength, geometric details or discontinuities with irregular cross sections and curvatures. Despite their limited number in relation to regions of mild material homogeneity, they generate nonseparable wave fronts which require extremely small grid densities. This significant hindrance, when encountered in subwavelength areas, can be efficiently ameliorated by a hybrid time-domain formulation [1] combining a multimodal FDTD-based algorithm of controllable resolution with a PSTD technique. Let us assume the discontinuity of Figure 6.1, spanning over an angle

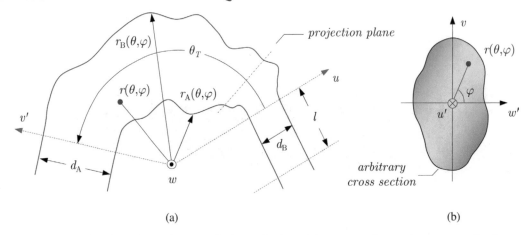

FIGURE 6.1: A 3-D arbitrarily curved discontinuity with constantly varying geometric features. (a) Transverse plane view and (b) cross-sectional plane view.

θ_T with inner and outer mean radii $r_A(\theta,\varphi)$ and $r_B(\theta,\varphi)$. The key issue is the systematic separation of *all* propagating modes at κ preselected transverse planes, which satisfy the necessary physical conditions. So, each component f can be expressed via infinite series as $r_M(\theta,\varphi) = r_B(\theta,\varphi) - r_A(\theta,\varphi)$, ζ_κ are the respective eigenfunctions and F_κ depict amplitude coefficients.

$$f(r,\theta,\varphi) = \sum_\kappa \zeta_\kappa(r,\theta,\varphi)F_\kappa(r,\theta,\varphi), \tag{6.1}$$

where

$$\zeta_\kappa(r,\theta,\varphi) = M_\kappa \cos\left[\kappa\pi\left(\frac{r - r_A(\theta,\varphi)}{r_M(\theta,\varphi)}\right)\right], \tag{6.2}$$

$$M_\kappa = \sqrt{\frac{2 - \delta_{\kappa 0}}{r_M(\theta,\varphi)}} \quad \text{for} \quad \delta_{\kappa\lambda} = \int_{r_A}^{r_A + d_A} \zeta_\kappa(r,\theta,\varphi)\zeta_\lambda(r,\theta,\varphi)\mathrm{d}r, \tag{6.3}$$

An important quality of (6.1)–(6.3) is their conformal character, which enables the rigorous discretization of the above oddities through the careful selection of $r_A(\theta,\varphi)$ and $r_B(\theta,\varphi)$. Two adequately smooth functions that fulfill this goal, with $b_{\theta,\varphi} = (\theta - \varphi)/\theta_T$, $d = d_B - d_A$, $d_B = 1.5d_A$, and $\tau = \omega/c$, are

$$r_A(\theta,\varphi) = d\left(b_{\theta,\varphi} - 1.25\right) + d_A - 0.35d_B, \tag{6.4}$$

$$r_{\mathrm{B}}(\theta, \varphi) = -d\left(b_{\theta,\varphi} - 1.5\right) + 0.1d_{\mathrm{A}} + 0.25d_{\mathrm{B}} \tag{6.5}$$

Having conducted the appropriate decoupling of the strenuous modes, the parametric technique projects Maxwell's equations on the prefixed κ planes. Consequently, a set of differential equations is extracted whose solutions act as transverse intermediate excitation surfaces in the discontinuity. Such a mechanism is proven very precise and the most substantial: it preserves lattice duality even close to the geometric peculiarity, contrary to common methods that are unable to provide sufficient treatment. In this way and through a matrix notation, Ampère's and Faraday's laws become

$$\varepsilon \frac{\partial \mathbf{E}}{\partial t} = \tau^{-1} \left(\mathbf{K}^{\mathrm{A}} + \mathbf{K}^{\mathrm{B}} \mathbf{K}^{\mathrm{C}}\right) \nabla \times \mathbf{H} - \mathbf{K}^{\mathrm{D}} \mathbf{E}, \tag{6.6}$$

$$\mu \frac{\partial \mathbf{H}}{\partial t} = -\tau \mathbf{K}^{\mathrm{C}} \nabla \times \mathbf{E} + \left(\mathbf{K}^{\mathrm{E}} - \mathbf{K}^{\mathrm{D}}\right) \mathbf{H}, \tag{6.7}$$

with $\mathbf{E}$ and $\mathbf{H}$ as the electric and magnetic intensities defined at the general coordinates (u, v, w). The elements of $\mathbf{K}^i$ impedance matrices, for $i = \mathrm{A}, \ldots, \mathrm{E}$, give the fundamental details of the structure and are derived, via all field continuity boundary constraints, as

$$K_{\kappa\lambda}^{\mathrm{A}} = (-1)^{\kappa+\lambda} \kappa^2/(\kappa^2 + \lambda^2), \qquad K_{\kappa\lambda}^{\mathrm{B}} = (\tau^2 - 1)\delta_{\kappa\lambda}, \tag{6.8}$$

$$K_{\kappa\lambda}^{\mathrm{C}} = r_{\mathrm{A}}(\theta, \varphi) + d/2, \qquad K_{\kappa\lambda}^{\mathrm{B}} = r_{\mathrm{M}}(\theta, \varphi)\delta_{\kappa\lambda} - d/4, \tag{6.9}$$

$$K_{\kappa\lambda}^{\mathrm{E}} = (-1)^{\kappa+\lambda} M_\kappa M_\lambda r_{\mathrm{B}}(\theta, \varphi)/(\kappa^2 + \lambda^2) \tag{6.10}$$

To approximate the spatial derivatives of (6.6) and (6.7), the Nth-order finite-difference operator of

$$\left.\frac{\partial f}{\partial \xi}\right|_l^n = \frac{2N+1}{\Delta\xi} \sum_{s=-(2N-1)}^{2N-1} a_{|s|} f|_s^n = \mathbf{D}_{\mathrm{A}}^N\left[f|_l^n\right] \cdot \mathbf{D}_{\mathrm{B}}^N\left[f|_l^n\right], \tag{6.11}$$

is constructed [2, 3], where $\xi \in (u, v, w)$, $l \in (i, j, k)$, $\Delta\xi$ is the spatial increment, s indicates specific grid points, and $a_{|s|}$ are weighting parameters. Furthermore, approximators $\mathbf{D}_{\mathrm{A}}[.]$ and $\mathbf{D}_{\mathrm{B}}[.]$, defined as

$$\mathbf{D}_{\mathrm{A}}^N\left[f|_l^n\right] = \sum_{s=1}^{N} \frac{(\Delta\xi)^{2s}}{(2s+1)!2^{2s}} \left(f|_{l+s}^n - f|_{l-s+1}^n\right), \tag{6.12}$$

$$\mathbf{D}_{\mathrm{B}}^N\left[f|_l^n\right] = \sum_{s=1}^{N} \frac{(\Delta\xi)^{2s+1}}{(s+2)!2^{s-1}} \left(f|_{l+s-1/2}^n - f|_{l-s+1/2}^n\right), \tag{6.13}$$

launch additional nodal patterns that optimize mesh density and improve the convergence of the schemes.

Proceeding to the second part of the hybridization, a curvilinear Fourier-Chebyshev PSTD interpolation [4–6] is devised for sections of homogeneous media and periodical features. Because they are the most frequently encountered in the domain, their examination via the prior approach is expected to radically accelerate the total simulation. This is achieved by evaluating spatial derivatives in terms of

$$\left.\frac{\partial f}{\partial \xi}\right|_\Xi^n = \frac{2\pi}{L_\xi \Delta \xi} \mathcal{F}^{-1}\left\{ jk_\xi \mathcal{F}\left\{ f|_{i,j,k}^n \right\} \right\}, \tag{6.14}$$

with $\mathcal{F}\{.\}$ and $\mathcal{F}^{-1}\{.\}$ the Fourier and inverse Fourier transform, L_ξ the node number toward ξ-axis, and k_ξ the transformed wavevector component. Moreover, Ξ describes all ξ coordinates along a straight-line cut through the transverse plane of the other two coordinates. In fact, (6.14) represents differentiation by the following trigonometric Chebyshev-type polynomial of

$$\left.\frac{\partial f}{\partial \xi}\right|_\Xi^n = \frac{1}{L_\xi} \sum_{m=-L_\xi/2}^{m=L_\xi/2-1} j\frac{2\pi m}{L_\xi} \tilde{f}(m)|_\Xi^n e^{j2\pi m\Xi/L_\xi} \tag{6.15}$$

with the tilde over $f(m)$ denoting the Fourier series of f and $j^2 = -1$. Observe that all field values are placed at certain *nonuniform* grid points $\psi_i = \pi i/L_\xi$, for $i = 0, 1, \ldots, 2L_\xi - 1$. This fact permits space partition into collocated elements, which conform to the boundaries of media distributions. Next and by means of a curvilinear coordinate transformation, each element is mapped into a cube, whereas, finally, all field quantities adjacent to internal interfaces, are properly corrected to resolve the physical jump conditions.

A prominent merit of the hybrid FDTD/PSTD forms [4] is that temporal derivatives are computed through

$$\left.\frac{\partial f}{\partial t}\right|_l^n = \frac{17}{12\Delta t}\left(f|_l^{n+1/2} - f|_l^{n-1/2} \right) + O(\Delta t^2), \tag{6.16}$$

thus facilitating their completely independent temporal update. A closer inspection of (6.11)–(6.13) and (6.15), unveils their competence in manipulating frequency-dependent layouts due to the direct tensor-oriented classification of constitutive parameters. Hence, dispersion errors are considerably minimized and wideband investigations are straightforward to accomplish.

6.3   DOMAIN DECOMPOSITION AND INTERFACE BOUNDARY CONDITIONS

A fundamental point in modern large-scale EMC setups is the periodical repetition of identical geometric regions. To evade this structural redundancy, a 3-D domain decomposition method is discussed, according to which the computational space Ω is divided into M disjoint subdomains Ω_i [9–13]. Assume, without loss of generality, two nonoverlapping subdomains ($i = 1, 2$), where $\mathbf{E}$ and $\mathbf{H}$ vectors are independently computed by the aforesaid hybrid schemes. Special attention is drawn on the interface, at the edges of which versatile field continuity conditions are enforced. Hence, $\mathbf{E}_1$ and $\mathbf{H}_2$ (dually $\mathbf{E}_2$ and $\mathbf{H}_1$) vectors, therein, are related by

$$\hat{\mathbf{n}}_1 \times \frac{1}{\mu_{r,1}} \nabla \times \mathbf{E}_1 = -\mu_0 \hat{\mathbf{n}}_2 \times \frac{\partial \mathbf{H}_2}{\partial t}, \tag{6.17a}$$

$$\hat{\mathbf{n}}_2 \times \frac{1}{\varepsilon_{r,2}} \nabla \times \mathbf{H}_2 = \varepsilon_0 \hat{\mathbf{n}}_1 \times \frac{\partial \mathbf{E}_1}{\partial t} + \sigma_1 \frac{1}{\varepsilon_{r,1}} \hat{\mathbf{n}}_1 \times \mathbf{E}_1 \tag{6.17b}$$

Furthermore, the equivalent surface current concept is introduced as

$$\mathbf{J}_{s,i} = \hat{\mathbf{n}}_i \times \frac{1}{\mu_{r,i}} \nabla \times \mathbf{E}_i \quad \text{for} \quad i = 1, 2, \tag{6.18}$$

and $\hat{\mathbf{n}}_i$ outward normal unit vectors. Enhancing (6.6) and (6.7) by Lagrange multipliers $\gamma_{1,2}$ via (6.17), gives

$$\frac{1}{\mu_{r,1}} \nabla \times \mathbf{E}_1 = -\frac{\partial \mathbf{B}_1}{\partial t} + \gamma_1 \left\{ \hat{\mathbf{n}}_2 \times \frac{1}{\mu_{r,2}} \nabla \times \mathbf{E}_2 + \nabla \times \mathbf{H}_2 \right\}, \tag{6.19a}$$

$$\frac{1}{\varepsilon_{r,2}} \nabla \times \mathbf{H}_2 = \frac{\partial \mathbf{D}_2}{\partial t} + \sigma_2 \mathbf{E}_2 + \gamma_2 \left\{ \nabla \times \mathbf{E}_1 + \hat{\mathbf{n}}_1 \times \frac{1}{\varepsilon_{r,1}} \nabla \times \mathbf{E}_1 \right\} \tag{6.19b}$$

at the interface. Exploiting the particular features of the most frequently encountered EMC applications, the maximum number of subdomains is confined to 14, i.e., two central, six face, and six edge ones. Besides, in areas of mild field changes, (6.6)–(6.16) may combined with surface impedance conditions [14].

Nonetheless, apart from the potential of decomposing the computational domain, the consistent interconnection of FDTD and PSTD areas in the hybrid technique of the previous section must be also an issue of serious study. As a matter of fact, an advantageous treatment may be obtained through the fluxes across the boundaries. In particular within each region, the fields are updated independently and at the end of the time-step, waves are separated into two parts: a normally incident to the boundary and one propagating away from it. For example, the corrected tangential electric components are given by

$$\hat{\mathbf{n}} \times \mathbf{E} = \hat{\mathbf{n}} \times \frac{(Z\mathbf{E} - \hat{\mathbf{n}} \times \mathbf{H})^- + (Z\mathbf{E} + \hat{\mathbf{n}} \times \mathbf{H})^+}{Y^- + Y^+}, \tag{6.20}$$

where $\pm$ indicates the left/right region with respect to the boundary and Z and Y are the medium's impedance and admittance, respectively. The theoretical performance of the FDTD/PSTD is explored by the associated error between the numerical k_{num} and exact wavenumber. Therefore, the dispersion relation is

$$k_{num}(\omega) \approx \omega \left\{ 1 - \frac{7(\Delta t)^{N+5}}{539} - \frac{5(\Delta t + 1)^{3N+1}}{814} + O(\Delta t)^{N+2} \right\} \qquad (6.21)$$

To display the notable accuracy improvement, Figure 6.2 shows the normalized phase velocity and the maximum L_2 error norm as a function of various factors at a spectrum of 12 GHz and compares the results with two characteristic higher-order FDTD schemes [7]. In addition, Table 6.1 provides

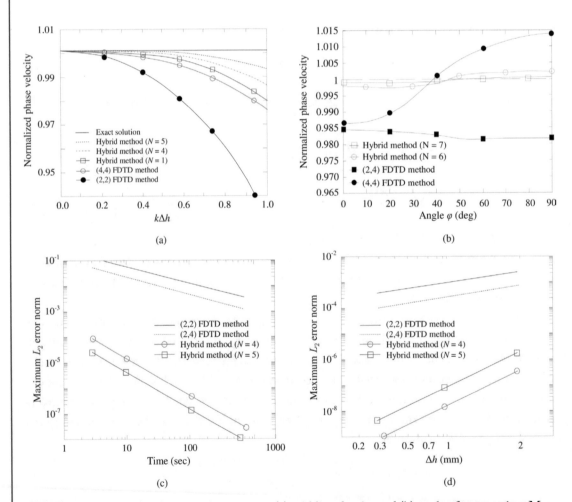

FIGURE 6.2: Normalized phase velocity versus (a) gridding density and (b) angle of propagation. Maximum L_2 error norm versus (c) time and (d) spatial increment.

METHOD	GRID DENSITY	CONVERGENCE	MAX. DISPERSION
TABLE 6.1: Comparison of convergence rate and maximum dispersion			
(2,2) FDTD	1/45	1.8342	8.921×10^{-3}
(4,4) FDTD	1/25	3.6184	3.745×10^{-4}
Hybrid ($N = 3$)	1/15	3.2817	2.304×10^{-8}
Hybrid ($N = 5$)	1/10	4.9067	3.952×10^{-11}
Hybrid ($N = 7$)	1/8	7.0123	5.641×10^{-13}

the convergence rate and lattice error of diverse realizations. The superior multifrequency performance of the new method over conventional ones is directly discernible.

6.4   INDICATIVE APPLICATION

To comprehend the merits of the hybrid algorithm and its assistance in the treatment of large-scale EMC problems, consider the dual reflector anechoic chamber of Figure 6.3 with a width of 7.2 m,

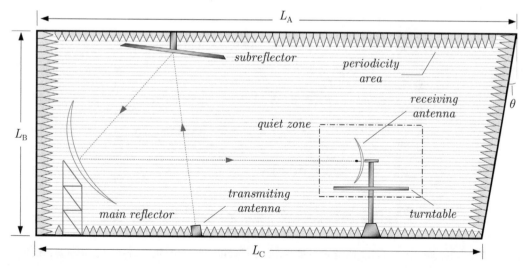

FIGURE 6.3: A dual reflector anechoic test chamber facility with a θ-inclined side wall.

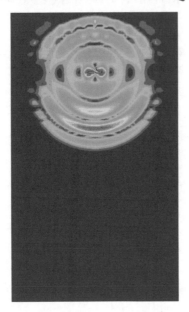 

FIGURE 6.4: Snapshots of the electric field intensity magnitude at a transverse cut of the dual reflector chamber quiet zone.

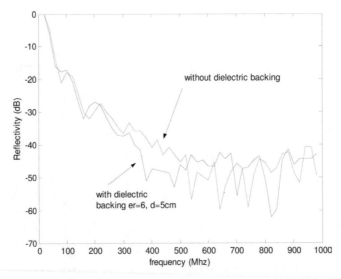

FIGURE 6.5: Reflectivity of the pyramidal absorber with and without the presence of its lossy dielectric backing.

L_A = 12.8 m, L_B = 5.6 m, and L_C = 11.4 m. As discussed in Chapter 5, in the specific facility, the subreflector (diameter 2.36 m) drives the excitation on the main reflector (diameter 4.4 m), whereas the inclination of the right wall is θ = 8.5°. The chamber is lined with 1.83 m tall, 25% carbon-loaded pyramids, backed by 7.6-mm ferrite tiles and properly tuned 5-cm-thick lossy dielectric (σ = 2.12 S/m, ε_r = 6) backing. The above setup leads to a 184 × 176 × 92 decomposed-domain lattice, where the multimodal FDTD method models areas of abrupt field fluctuation or high details and the PSTD technique handles all periodic details, such as the absorber linings.

For the preceding arrangement, Figure 6.4 presents two snapshots of the electric field intensity magnitude at a transverse cut of the chamber's quiet zone, before and after the illumination of the equipment under test. Apparently, the pyramidal lining works satisfactorily, because it manages to create a relatively uniform electromagnetic environment. The ripples at the edges of the zone

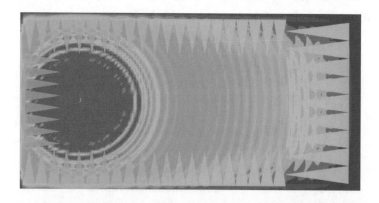

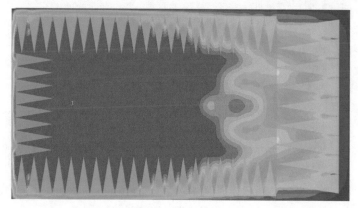

FIGURE 6.6: Snapshots of the electric field intensity magnitude at a transverse cut of the semianechoic chamber.

indicate its outer limits as well as the efficiency of the absorber. The latter quality, however, can be also verified through Figure 6.5, which illustrates the reflectivity of the pyramids with and without their lossy dielectric backing. As observed, in the range of 300–900 MHz, where most of emission and immunity measurements take place, the impact of the dielectric is rather important. It should be mentioned that if the particular facility was simulated only by means of the FDTD approach, the overall computational CPU and memory burden would have increased at least 65%.

Let us next investigate a rectangular $10.4 \times 6.4 \times 5.6$ m semianechoic chamber. The absorbing wall, behind the receiving antenna in the quiet zone, is located 3 m away from the transmitting one. The facility is equipped with an optimized set of hybrid 0.61 m tall, 27% carbon-loaded urethane absorbers, which are backed by a 5.2-mm-thick ferrite tile with a gap size of 0.2 mm and a 4-cm dielectric layer of $\varepsilon_r = 5.2$. Figure 6.6 depicts two different snapshots of the electric field intensity magnitude at a transverse cut of the chamber, whereas Figure 6.7 proves the capabilities of the combined FDTD/PSTD scheme by predicting the suitability of the chamber for EMI measurements at the low-frequency spectrum (normalized site attenuation between the ± 4 dB limits), unlike the misleading simulation of the conventional FDTD schemes.

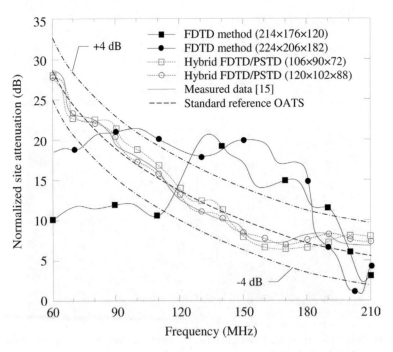

FIGURE 6.7: Normalized site attenuation of the rectangular semianechoic chamber.

6.5   ESD ASPECTS IN EMC ANALYSIS

ESD is a physical phenomenon that causes a transfer of electric charge between bodies of different electrostatic potential in proximity or through direct contact. This mechanism produces electromagnetic interference and can be regarded as a high-amplitude pulse that may cause noise in measuring instruments, permanent damage, latent failures, disruption in the functional operation of EMC devices, or even electric shocks to a person involved [15–19]. These consequences are mainly attributed to the rise time, amplitude, and duration of the current waveform. There are many variations in the charging process. A common situation is one where a person walks on a carpet. With each step, the person loses or gains electrons from the body to the fabric. Friction between the person's clothing and a chair can also produce an exchange of charge, often in the thousands of volts range. A conducting carpet provides no protection unless the operator is adequately earthed to it. Another typical ESD event can occur from static electricity generated when two media of diverse dielectric constitutive parameters rub against each other [20–25], whereas lightning could be also a cause [26–27]. Alternatively, charging of a material may be created via heating or through contact with a charged body. Moreover, media with a resistivity of more than 10^9 Ω are likely to develop electrostatic potentials, which are not easily discharged through leakage because of the high material insulation resistance.

The problem of static electricity accumulation and subsequent discharges becomes more relevant for uncontrolled environments. Actually, an object with a large charge density will seek the first available opportunity to discharge and reestablish its electric balance. This may simply happen by a gradual bleed of the charge via a moderately conducting path. Considering the widespread application of equipment and systems in everyday life, it is straightforward to recognize that their proper operation is prone to degradation, because they are subjected to electromagnetic fields whenever a discharge takes place from people standing nearby. To this extent, environmental and installation conditions can play a role of paramount significance. Not to mention that generation of ESD is favored by the combination of various fabrics and a dry atmosphere (low relative humidity). For these reasons, it is practically classified into two primary categories: human (direct) and furniture (air). The former is characterized by a fast rise of current, approximately 1 ns, up to a peak of 10 A, followed by a decay back to zero, whereas the latter exhibits a slower (gradual) rise of current to a peak of 40 A, accompanied by damped oscillations. It is therefore indisputable that ESD issues should not be neglected during the design process of any EMC component, because they can easily spoil any promising attempt.

6.6   NUMERICAL VERIFICATION

To comprehend the inherent attributes of an ESD event, consider the arrangement of Figure 6.8, which comprises two cylindrical cavities—connected via a 50-Ω coaxial cable—inside a shielded box.

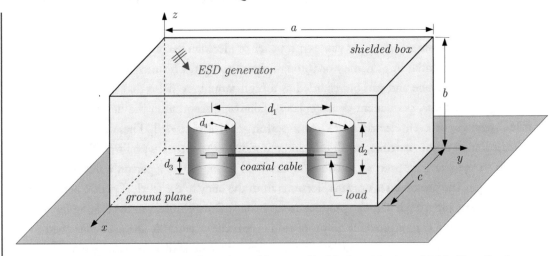

FIGURE 6.8: A characteristic configuration with two cylindrical cavities in a shielded box for the measurement of the ESD radiated field.

The dimensions of the structure are: $a = 160$ cm, $b = 48$ cm, $c = 55$ cm, $d_1 = 100$ cm, $d_2 = 15.8$ cm, $d_3 = 8.5$ cm, and $d_4 = 6.5$ cm. This arrangement prevents possible pickup of radiated fields by the coaxial cable. In fact, the ESD event takes place on the external side of the shielded enclosure connected to the ground plane.

In a real measurement system, if noise is nontrivial, the cable is additionally reinforced by ferrite of other absorber material in an effort to enhance its isolation effectiveness. The aforementioned

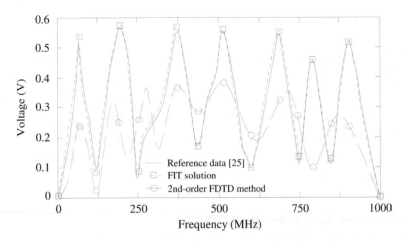

FIGURE 6.9: Induced voltage magnitude versus frequency at a distance of 35 cm from the ESD event.

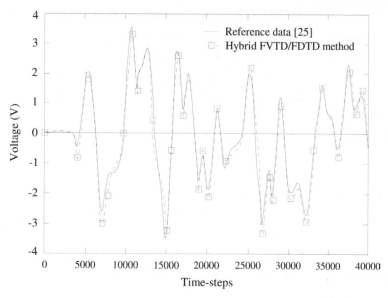

FIGURE 6.10: Induced voltage versus time-steps at a distance of 45 cm from the ESD event.

problem is simulated by means of the hybrid FVTD/FDTD method in a $124 \times 176 \times 82$ grid, where the former technique models the curved surfaces of the cavities as well as the region in the vicinity of the cable, and the latter discretizes the rest of the space. Figure 6.9 gives the magnitude of the induced voltage versus frequency at a 35-cm distance from the ESD event, revealing the promising accuracy of the hybrid approach in contrast to the rather poor results of the traditional FDTD implementation. Similar deduction can be extracted from Figure 6.10, which displays the induced voltage as a function of time at a distance of 44 cm from the ESD setup. Evidently, the specific application could be also solved through other combined schemes, such as the FETD/FDTD or the FIT/FDTD ones.

REFERENCES

1. N. V. Kantartzis and C. S. Antonopoulos, "Design of hybrid time-domain schemes with optimal gridding density and material-interface sensitivity for large-scale EMC problems," *IEEE Trans. Magn.*, vol. 44, no. 6, pp. 1462–1465, June 2008.

2. N. V. Kantartzis and T. D. Tsiboukis, "Higher-order non-standard schemes in generalised curvilinear coordinates: A systematic strategy for advanced numerical modeling and consistent topological perspectives," *ICS Newsl.*, vol. 9, no. 3, pp. 5–14, 2002.

3. N. V. Kantartzis, "A generalised higher-order FDTD-PML algorithm for the enhanced analysis of 3-D waveguiding EMC structures in curvilinear coordinates," *IEEE Proc. Microwave, Antennas Propagat.*, vol. 150, no. 5, pp. 351–359, Oct. 2003. doi:10.1049/ip-map:20030269

4. Q. H. Liu, "The PSTD algorithm: A time-domain method requiring only two cells per wavelength," *Microwave Opt. Technol. Lett.*, vol. 15, no. 3, pp. 158–165, 1997. doi:10.1002/(SICI)1098-2760(19970620)15:3<158::AID-MOP11>3.0.CO;2-3

5. Q. H. Liu, "Large-scale simulations of electromagnetic and acoustic measurements using the pseudospectral time-domain (PSTD) algorithm," *IEEE Trans. Geosci. Remote Sensing*, vol. 37, pp. 917–926, 1999.

6. Q. H. Liu and G. Zhao, "Advances in PSTD Techniques," in *Computational Electromagnetics: The Finite-Difference Time-Domain Method*, A. Taflove and S. C. Hagness, Norwood, MA: Artech House, 2005, chap. 17, pp. 847–882.

7. N. V. Kantartzis and T. D. Tsiboukis, *Higher-Order FDTD Schemes for Waveguide and Antenna Structures*, San Rafael, CA: Morgan & Claypool Publishers, 2006. doi:10.2200/S00018ED1V01Y200604CEM003

8. M. Tsumura, Y. Baba, N. Nagaoka, and A. Ametani, "FDTD simulation of a horizontal grounding electrode and modeling of its equivalent circuit," *IEEE Trans. Electromagn. Compat.*, vol. 48, no. 4, pp. 817–825, Nov. 2006.

9. C. Farhat, M. Lesoinne, P. Le Tallec, K. Pierson, and D. Rixen, "FETI-DP: a dual-primal unified FETI method—Part I: A faster alternative to the two-level FETI method," *Int. J. Numer. Methods Eng.*, vol. 50, pp. 1523–1544, Mar. 2001. doi:10.1002/nme.76

10. S. Lee, M. Vouvakis, and J.-F. Lee, "A non-overlapping domain decomposition method with non matching grids for modeling large finite arrays," *J. Comp. Phys.*, vol. 203, pp. 1–21, 2005. doi:10.1016/j.jcp.2004.08.004

11. Y. Li and J.-M. Jin, "A vector dual-primal finite element tearing interconnecting method for solving 3-D large-scale electromagnetic problems," *IEEE Trans. Antennas Propagat.*, vol. 54, no. 10, pp. 3000–3009, Oct. 2006. doi:10.1109/TAP.2006.882191

12. K. Zhao, V. Rawat, S.-C. Lee, and J.-F. Lee, "A domain decomposition method with nonconformal meshes for finite periodic and semi-periodic structures," *IEEE Trans. Antennas Propagat.*, vol. 55, no. 9, pp. 2559–2570, Sept. 2007.

13. Y. Li and J.-M. Jin, "A new dual-primal domain decomposition approach for finite element simulation of 3-D large-scale electromagnetic problems," *IEEE Trans. Antennas Propagat.*, vol. 55, no. 10, pp. 2803–2810, Oct. 2007.

14. M. K. Kärkkäinen, "FDTD surface impedance model for coated conductors," *IEEE Trans. Electromagn. Compat.*, vol. 46, no. 2, pp. 222–233, May 2003.

15. L. H. Hemming, *Electromagnetic Anechoic Chambers: A Fundamental Design and Specification Guide*. New York: IEEE Press and Wiley Interscience, 2002.

16. R. Chundru, D. Pommerenke, K. Wang, T. Van Doren, F. P. Centola, and J. S. Huang, "Characterization of human metal ESD reference discharge event and correlation of generator pa-

rameters to failure levels—Part I: Reference event," *IEEE Trans. Electromagn. Compat.*, vol. 46, no. 4, pp. 498–503, Nov. 2004.

17. K. Wang, D. Pommerenke, R. Chundru, T. Van Doren, F. P. Centola, and J. S. Huang, "Characterization of human metal ESD reference discharge event and correlation of generator parameters to failure levels—Part II: Correlation of generator parameters to failure levels," *IEEE Trans. Electromagn. Compat.*, vol. 46, no. 4, pp. 504–510, Nov. 2004. doi:10.1109/TEMC.2004.837688

18. Z. Yuan, T. Li, J. He, S. Chen, and R. Zeng, "New mathematical descriptions of ESD current waveform based on the polynomial of pulse function," *IEEE Trans. Electromagn. Compat.*, vol. 48, no. 3, pp. 589–591, Aug. 2006. doi:10.1109/TEMC.2006.877786

19. S. Caniggia and F. Maradei, "Numerical prediction measurement of ESD radiated fields by free-space field sensors," *IEEE Trans. Electromagn. Compat.*, vol. 49, no. 3, pp. 494–503, Aug. 2007.

20. A. Gavrilakis, A. P. Duffy, K. G. Hodge, and J. Willis, "Partial capacitance calculation for shielded twisted pair cables," *IEEE Trans. Electromagn. Compat.*, vol. 46, no. 2, pp. 299–301, May 2004. doi:10.1109/TEMC.2004.826883

21. H. Haase, T. Steinmetz, and J. Nitsch, "New propagation models for electromagnetic waves along uniform and nonuniform cables," *IEEE Trans. Electromagn. Compat.*, vol. 46, no. 3, pp. 345–352, Aug. 2004. doi:10.1109/TEMC.2004.831829

22. C. Buccela, M. Feliziani, and G. Manzi, "Detection and localization of defects in shielded cables by time-domain measurements with UWB pulse injection and clean algorithm processing," *IEEE Trans. Electromagn. Compat.*, vol. 46, no. 4, pp. 597–605, Nov. 2004. doi:10.1109/TEMC.2004.837842

23. G. C. Christoforidis, D. P. Ladridis, and P. S. Dokopoulos, "Inductive interference on pipelines buried in multilayer soil due to magnetic fields from nearby faulted power lines," *IEEE Trans. Electromagn. Compat.*, vol. 47, no. 2, pp. 254–262, May 2005. doi:10.1109/TEMC.2005.847399

24. M. Paolone, and E. Petrache, "Lightning induced disturbances in buried cables—Part I: Experimental and model validation," *IEEE Trans. Electromagn. Compat.*, vol. 47, no. 3, pp. 509–520, Aug. 2005. doi:10.1109/TEMC.2005.853163

25. G. Antonini and A. Orlandi, "Efficient transient analysis of long lossy shielded cables," *IEEE Trans. Electromagn. Compat.*, vol. 48, no. 1, pp. 42–56, Feb. 2006. doi:10.1109/TEMC.2006.870767

26. J.-P. Bérenger, "Long range propagation of lightning pulses using the FDTD method," *IEEE Trans. Electromagn. Compat.*, vol. 47, no. 4, pp. 1008–1011, Nov. 2005. doi:10.1109/TEMC.2005.858747

27. E. Petrache, F. Radici, M. Paolone, C. A. Nucci, V. A. Rakov, and M. A. Uman, "Lightning induced disturbances in buried cables—Part I: Theory," *IEEE Trans. Electromagn. Compat.*, vol. 47, no. 3, pp. 498–508, Aug. 2005. doi:10.1109/TEMC.2005.853161

CHAPTER 7

Contemporary Material Modeling in EMC Applications

7.1 INTRODUCTION

It is well acknowledged that during the past decade EMC research has attempted to expand its frontiers to novel and more challenging horizons, which are expected to provide pivotal answers in several open problems regarding the design of optimized devices. Toward this goal, the investigation of modern materials, not necessarily encountered in nature, can grant intuitive solutions and lay the foundations for some really advanced configurations. Being complex in their composition and internal structure, these modern materials require a meticulous study of their principal properties that often exposes the inadequacies of existing techniques. Even the recent numerical algorithms are sometimes incapable of manipulating a certain part of their interactions. As a consequence, complementary formulations and enhanced schemes should be developed, whereas the discretization strategies must incorporate extra degrees of freedom for the description of the new profiles.

Consenting to the above considerations, the present chapter explores the role of diverse contemporary media in the EMC practice via different time-domain methods. The first category is the family of metamaterials that permit the simultaneous adjustment of their media parameters to negative values and thus accomplish a unique performance. Next, analysis moves on to the general class of bi-isotropic substances and chiral arrangements in an effort to produce viable setups of highly efficient absorber linings or mounting substrates. These interesting assets are then applied to patch antennas belonging to the area of nanotechnology for the extraction of possible dimension thresholds and polarization patterns. The chapter closes with the simulation of some rod-based photonic crystal layouts in the form of waveguides and power splitters constructed by the systematic insertion or removal of specific elements from the body of the crystal.

7.2 DOUBLE-NEGATIVE METAMATERIALS

One of the most significant innovations in modern electromagnetics is the advent of *double-negative (DNG)* or *left-handed metamaterials*, whose promising properties have recently gained remarkable reputation [1–5]. In a DNG medium, the concurrent tuning of both constitutive parameters to negative real values leads to a negative refractive index and other features worth mentioning that are not accessible in nature, such as the inverted Snell law, Doppler shift, or Cherenkov radiation. Their initial conception dates back in 1967, when primary research theoretically predicted that a planar DNG slab can perfectly focus electromagnetic waves emanating from a point source. However, this far-sighted breakthrough remained inactive until the pioneering achievements of Pendry et al. [2], which triggered the construction of the first artificial DNG substance with thin metallic wires and the so-called *split-ring resonators (SRRs)*. Since then, the scientific community has delved into the metamaterial fundamental concepts to fabricate practicable structures or numerically explore their particular characteristics [6–30]. Devoted mainly to the challenging EMC perspective of DNG media, this section provides a brief record of their theoretical background along with several modern applications.

7.2.1   Theoretical Formulation

Let us consider an infinite DNG slab of width d that is parallel to the $x = 0$ plane of a Cartesian coordinate system. On the left-hand side, at a distance d_s, the slab is illuminated by an x-plane magnetic current source, as depicted in Figure 7.1. When losses are neglected, such a structure, with $\varepsilon_r = \mu_r = -1$, can focus electromagnetic waves, generated by the prior source, on the right-hand side plane at a distance $d_f = d - d_s$. However, in most realistic problems, losses and deviations from the preceding ideal case are inevitable, and, under certain circumstances, their impact may be proven crucial.

    The analysis concentrates, without loss of generality, on the 2-D z-polarized transverse electric (TE) case. For notational simplicity, the convention $E(x, y)$ for the sole electric field component

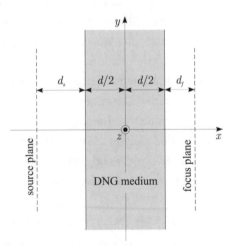

FIGURE 7.1: Geometry of an infinite lossy DNG slab parallel to the $x = 0$ plane.

along the z-axis is used. The transfer function from the $x = 0$ source plane to an arbitrary x-plane on the slab's right-hand side—hereafter designated as the observation plane—for a plane wave $e^{-jk_x x} e^{-jk_y y}$ is

$$T(k_y) = \frac{4\zeta e^{-jk_x(x-d)}}{(\zeta + 1)^2 e^{jk'_x d} - (\zeta - 1)^2 e^{-jk'_x d}}, \tag{7.1}$$

where $\zeta = \mu k_x / (\mu_0 k'_x)$, k_y is the y-component of the wavevector, and k_x, k'_x its x-directed counterparts in the vacuum-DNG space, which satisfy the dispersion relations $k_x^2 + k_y^2 = k_0^2$ and $k'^2_x + k_y^2 = k_0^2 n^2$, respectively. Notice that k_0 is the free-space wavenumber, whereas n is the medium's refractive index. If $E_{inc}(x = 0, y)$ is an arbitrary incident distribution on the source plane, the total field on the observation plane is given by

$$E_{tot}(x, y) = \frac{1}{2\pi} \int_{-\infty}^{+\infty} \check{E}_{inc}(k_y) T(k_y) e^{-jk_y y} dk_y, \tag{7.2}$$

with

$$\check{E}_{inc}(k_y) = \int_{-\infty}^{+\infty} E_{inc}(x = 0, y) e^{jk_y y} dy, \tag{7.3}$$

denoting a Fourier-expansion mode of the incident field with spatial frequency k_y. It is stressed that, for $|k_y| > k_0$ in the vacuum and $|k_y| > k_0|n|$ in the DNG region, evanescent waves are induced, otherwise propagating waves are generated. To this end, a sufficient choice for k_x, k'_x to preserve causality [25] is

$$k_x = k_0 \sqrt{1 - k_y^2 / k_0^2}, \qquad k'_x = k_0 n \sqrt{1 - k_y^2 / (k_0^2 n^2)}$$

for propagating waves and

$$k_x = -jk_0 \sqrt{k_y^2 / k_0^2 - 1} \qquad k'_x = -jk_0 n \sqrt{k_y^2 / (k_0^2 n^2) - 1}$$

for evanescent waves

Essentially, the most beneficial practical realization of a DNG medium comprises a network of periodically repeated rectangular or circular SRRs combined with metallic rods or strip wires, as depicted in Figure 7.2. The SRR grid, which in the millimeter spectrum presents a negative effective magnetic permeability, can be formulated by the *Lorentz (LO)* or *Drude (DR)* lossy model of

$$\mu_{eff}^{LO}(\omega) = \mu_0 \left(1 - \frac{F\omega^2}{\omega^2 - \omega_0^2 + j\Gamma_m \omega} \right) \quad \text{and} \quad \mu_{eff}^{DR}(\omega) = \mu_0 \left[1 - \frac{F\omega_{pm}^2}{\omega(\omega - j\Gamma_m)} \right], \tag{7.4}$$

where

$$\omega_0^2 = \frac{3l v_0^2}{\pi r^3 \ln(2d/g)} \quad \text{and} \quad F = \frac{\pi r^2}{b^2},$$

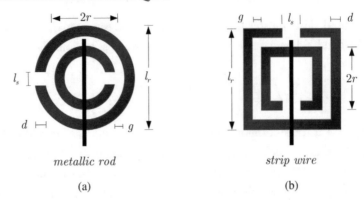

metallic rod

(a)

strip wire

(b)

FIGURE 7.2: (a) A circular SRR combined with a thin metallic rod and (b) a rectangular SRR combined with a metallic strip wire.

b is the grid constant, r is the inner ring radius, and d the width of the SRR. In addition, ω_{pm} denotes the plasma and $\Gamma_m = 2\rho l/r$ the damping frequency, with l the distance of adjacent planes, ρ the perimeter resistance, and υ_0 the light velocity. The double splits and ring gap, g, provide an ample capacitance which ensures that the element's resonant wavelength is always larger than the SRR diameter.

In contrast, the second DNG component, i.e., the set of thin rods or wires, behaves like a quasi-medium with a negative effective dielectric permittivity, described in the microwave regime by

$$\varepsilon_{\text{eff}}^{\text{LO}}(\omega) = \varepsilon_0 \left[1 - \frac{(\varepsilon_s - 1)\omega^2}{\omega^2 - \omega_0^2 + j\Gamma_e \omega} \right] \quad \text{and} \quad \varepsilon_{\text{eff}}^{\text{DR}}(\omega) = \varepsilon_0 \left[1 - \frac{(\varepsilon_s - 1)\omega_{pl}^2}{\omega(\omega - j\Gamma_e)} \right], \quad (7.5)$$

where ω_{pl} and Γ_e are the respective plasma and damping frequencies and ε_s the static dielectric constant. Even though both models in (7.4) are theoretically equivalent, the former is more suitable for the thorough study of losses, whereas the latter supports a broader bandwidth of concurrently negative constitutive parameters. This simply means that Drude solutions reach steady state faster than the Lorentz ones [31]. So, when a monochromatic plane wave travels in such a medium, its Poynting vector is antiparallel to the direction of its phase velocity.

7.2.2   A Generalized Time-Domain Methodology

Having derived the basic theoretical framework for the analysis of DNG media, our investigation proceeds to their behavior as a function of several physical and geometric parameters. To this end, an efficient algorithm for the evaluation of the cross correlation between the field at the source and any observation plane is developed. So, a rigorous tool that indicates the resemblance of electric field distributions at the prior planes is introduced via two different approaches [25]. The former computes the correlation function

$$\rho(x) = \frac{\left|\int_{-\infty}^{+\infty} E_{\text{inc}}(0,y) E_{\text{tot}}^*(x,y) dy\right|^2}{\int_{-\infty}^{+\infty} |E_{\text{inc}}(0,y)|^2 \, dy \, \int_{-\infty}^{+\infty} |E_{\text{tot}}(x,y)|^2 \, dy}, \tag{7.6}$$

with $E_{\text{tot}}^*(x, y)$ the conjugate of $E_{\text{tot}}(x,y)$, whereas the latter calculates the enhanced mean square error (mse)

$$\text{mse}_m(x) = \frac{\int_{-\infty}^{+\infty} \left| E_{\text{inc}}(0,y) - E_{\text{tot}}(x,y) e^{-j \arg\{E_{\text{tot}}(x,0)\}} \right|^2 dy}{\int_{-\infty}^{+\infty} |E_{\text{inc}}(0,y)|^2 \, dy}, \tag{7.7}$$

where the extra exponential in the nominator cancels any misleading phase terms generated by the vacuum-DNG interface. Observe that the closer $\rho(x)$ approaches unity or mse(x) zero, the stronger the resemblance is between the two fields. Hence, the observation plane on which $\rho(x)$ is maximized or mse(x) is minimized constitutes the desired focus plane. In this context, to accomplish fast and reliable design estimates, the DNG medium is examined via a rigorous dispersive FDTD algorithm. For the simulations, a Drude model—instead of the more laborious Lorentz one—is used to derive convenient update equations. Nevertheless, despite the possible model selection, results (as already mentioned) are anticipated to be the same, because the metamaterials are studied at a certain frequency where Re$\{n\} = -1$ or close to it. Recalling (7.4) and (7.5), the relative electric permittivity and magnetic permeability of a Drude material are expressed as

$$\varepsilon_r(\omega) = 1 + \frac{\omega_{pe}^2}{\omega(-\omega + j\Gamma_e)} \quad \text{and} \quad \mu_r(\omega) = 1 + \frac{\omega_p^2}{\omega(-\omega + j\Gamma_m)} \tag{7.8}$$

In most applications, these media are matched to free space, namely, their wave impedance equals to the free space one, so that a normally impinging wave has zero reflections. To satisfy this requirement, the relative electric permittivity and magnetic permeability must be equal, which yields $\omega_{pe} = \omega_{pm} = \omega_p$ and $\Gamma_e = \Gamma_m = \Gamma$. Also, if at the central frequency $\varepsilon_r(\omega_0) = \mu_r(\omega_0) = -1.0 - j\delta$, it is easy to extract that $\omega_p = \omega_0 \sqrt{2}$ and $\Gamma = \omega_0 \delta/2$. Therefore, the time-dependent Maxwell's curl equations in the DNG medium are

$$\nabla \times \mathbf{H} = \mathbf{J} + \varepsilon_0 \frac{\partial \mathbf{E}}{\partial t}, \qquad -\nabla \times \mathbf{E} = \mathbf{K}_s + \mathbf{K} + \mu_0 \frac{\partial \mathbf{H}}{\partial t}, \tag{7.9}$$

where $\mathbf{K}_s$ is the magnetic current source distribution. On the other hand, the constitutive relations for currents $\mathbf{J}$ and $\mathbf{K}$ are

$$\frac{\partial \mathbf{J}}{\partial t} + \Gamma \mathbf{J} = \omega_p^2 \varepsilon_0 \mathbf{E}, \qquad \frac{\partial \mathbf{K}}{\partial t} + \Gamma \mathbf{K} = \omega_p^2 \mu_0 \mathbf{H} \tag{7.10}$$

Selecting a TE-polarized field, it is deduced from (7.10) that only the J_z electric current component and their K_x, K_y magnetic current counterparts exist in our simulations. Essentially, the distributed source, used in (7.9), is a magnetic current sheet $\mathbf{K}(x, t) = f(x)g(t)\hat{\mathbf{z}}$, with $f(x)$ representing the spatial and $g(t)$ the temporal evolution. To this extent, the domain is discretized in a 2-D lattice with Δx and Δy as its spatial increments and Δt the temporal step. Apart from the $\mathbf{E}$ and $\mathbf{H}$ field quantities, which comply with the FDTD topology, J_z, K_x, and K_y are collocated with E_z at grid vertices. Consequently, if the vacuum-DNG interface coincides with the secondary grid lines, the boundary conditions—continuity of tangential magnetic field components—will be promptly fulfilled. Note that J_z are sampled at half-integer and K_x, K_y at integer time-steps. So, the FDTD equations for spatial and temporal derivatives are extracted via a central differencing process. For instance, the update formula for E_z is

$$E_z|_{i,j}^{n+1} = E_z|_{i,j}^{n} + \frac{\Delta t}{\Delta x}\left(H_y|_{i+1/2,j}^{n+1/2} - H_y|_{i-1/2,j}^{n+1/2}\right) - \frac{\Delta t}{\Delta y}\left(H_x|_{i,j+1/2}^{n+1/2} - H_x|_{i,j-1/2}^{n+1/2}\right) - \Delta t J_z|_{i,j}^{i+1/2} \quad (7.11)$$

7.2.3 Metamaterial-Based EMC Applications

In this paragraph, several time-domain numerical methodologies are used for the solution of real-world metamaterial-based EMC problems loaded with networks of SRRs and thin rods. The first structure is the two-port WR137 waveguide of Figure 7.3, which contains two parallel metamaterial planes for the enhancement of its functional spectrum. The circular SRR dimensions, as indicated in Figure 7.2, are: $l_r = 3.26$ mm, $r = 0.92$ mm, $l_s = 0.28$ mm, $g = 0.31$ mm, and $d = 0.24$ mm. The analysis is performed via the enhanced alternating-direction implicit (ADI)-FDTD method, incorporating a set of accurate spatial/temporal nonstandard operators (see Chapter 5). Table 7.1 summarizes the first resonance frequency for three different geometries regarding length and grid constant (<u>case A:</u> $l_w = 15.78$ mm, $a = 3.44$ mm; <u>case B:</u> $l_w = 23.92$ mm, $a = 4.54$ mm; <u>case C:</u> $l_w = 30.16$ mm, $a = 5.62$ mm). As detected, the ADI-FDTD method overwhelms the second-order FDTD one, yielding very

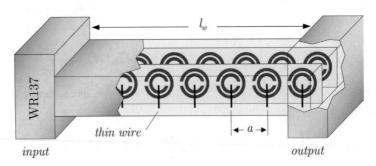

FIGURE 7.3: A WR137 two-port waveguide loaded with two parallel sheets of SRR-wire networks.

REFERENCE (GHz) [4]	METHOD (CFLN)	COMPUTED (GHz)	ERROR (%)	LATTICE (CELLS)	MAXIMUM DISPERSION
TABLE 7.1: First resonance frequency of the DNG-loaded waveguide					
Case A 5.8524	FDTD	5.6351	3.7134	$132 \times 246 \times 64$	1.962345×10^{-1}
	ADI-FDTD (24)	5.8517	0.0114	$50 \times 138 \times 30$	4.175862×10^{-11}
Case B 7.6231	FDTD	7.1804	5.8072	$140 \times 264 \times 68$	2.168563
	ADI-FDTD (28)	7.6208	0.0296	$56 \times 142 \times 32$	3.487901×10^{-10}
Case C 9.4875	FDTD	8.6939	8.3641	$146 \times 278 \times 74$	2.847105
	ADI-FDTD (32)	9.4836	0.0409	$62 \times 146 \times 34$	6.307194×10^{-10}

small (almost 10 orders lower) dispersion errors for fairly coarse curvilinear grids and large Courant-Friedrich-Levy number (CFLN).[1]

Proceeding to a more complex device, the four-port cross junction of Figure 7.4 is next examined. Its interior is loaded with two types of circular SRR-wire arrangements placed both transversely ($l_r = 3.84$ mm, $r = 0.98$ mm, $l_s = 0.32$ mm, $g = 0.38$ mm, $d = 0.29$ mm) and ($l_r = 2.56$ mm, $r = 0.72$ mm, $l_s = 0.16$ mm, $g = 0.24$ mm, $d = 0.19$ mm) along the structure's axis.

The junction with $l_A = l_D = 5.7$ mm, $l_B = 11.3$ mm, and $l_C = l_E = 2.6$ mm is simulated by means of the finite integration technique and the hybrid finite-volume time-domain (FVTD)/FDTD method, which leads to a mesh of $60 \times 154 \times 26$ cells. Figure 7.5 illustrates the S_{12}, S_{34} parameters and proves the efficiency of both algorithms in the modeling of curvilinear details, unlike the outcomes of the usual FDTD technique that needs a 85% finer lattice and still cannot reach sufficient levels of accuracy.

Apart from the traditional SRR-rod configurations, metamaterials can be engineered through alternative designs as well, like the spiral-inductor SRR element of Figure 7.6a. Placed according to the pattern shown in Figure 7.6b, these structures may be used for the fabrication of specialized EMC components. In this framework, the first application is a microstrip antenna mounted on a DNG substrate of the aforesaid kind. A set of typical dimensions is: $l_A = 8.6$ mm, $l_B = 9.1$ mm, $l_C = 5.2$ mm,

---

[1]The number in parenthesis at the second column of Table 7.1 indicates how large is the temporal increment of the ADI-FDTD technique as compared to the one dictated by the Courant stability condition.

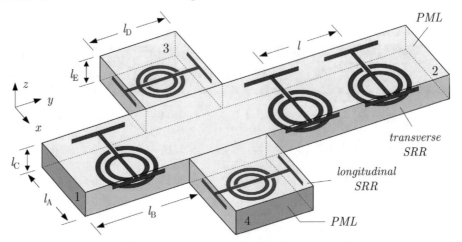

FIGURE 7.4: A DNG-loaded four-port junction with longitudinal and transverse SRR-wire layouts.

$l_D = 5.9$ mm, $l_s = 0.6$ mm, $h = 0.25$ mm, s = 0.23 mm, and $l = 17.4$ mm. Simulations are conducted in terms of the nonstandard (ND)-FDTD method and the characteristic impedance for a 5×5 and 10×10 periodic setup are given in Figure 7.7a. As easily deduced, the ND-FDTD algorithm exhibits a very good agreement with the reference data [5]. On the other hand, the second device is a PCB bandpass filter on a spiral-inductor SRR layer. Keeping the dimensions of the elements the same,

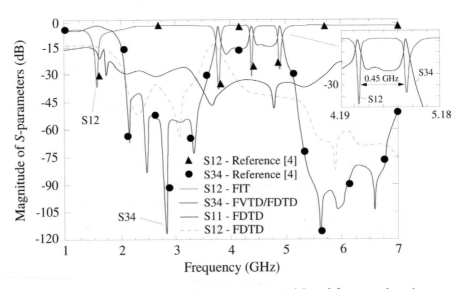

FIGURE 7.5: Calculation of S parameters for the metamaterial-based four-port junction.

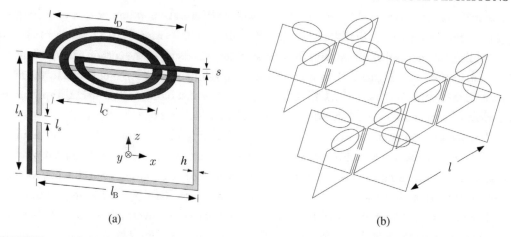

(a)   (b)

FIGURE 7.6: (a) An SRR-based spiral inductor. (b) A periodic spiral-inductor SRR antenna substrate.

Figure 7.7b depicts the transmitted and reflected power of the filter, as computed via the transmission-line modeling method. Note that the frequency band gaps in the figure correspond *to* the regions of unobstructed transmission, an issue that points out the distinct spectra of the filter and confirms its correct operation.

The impact of metamaterials in the fabrication of shielding surfaces with an advanced performance is, subsequently, investigated. Let us assume the 90° bend of Figure 7.8a (only a small portion

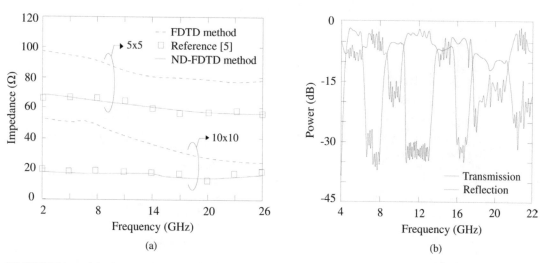

(a)   (b)

FIGURE 7.7: (a) Characteristic impedance of two array antennas with a spiral SRR substrate, and (b) transmitted and reflected power of a bandpass filter implemented through a spiral SRR layer.

is presented) that is constructed by means of rectangular SRRs and a continuous thin wire. By properly tuning the dimensions of the latter parts ($l_r = 3.72$ mm, $r = 0.98$ mm, $l_s = 0.32$ mm, $g = 0.31$ mm, $d = 0.24$ mm), the transmitted power and thus the shielding efficiency of the curtain can be adequately optimized. For example, Figure 7.8b indicates the accomplishment (solution is obtained by the finite-element time-domain method) of a 1-GHz band gap in the area of 8.5 to 10.7 GHz, implying that the surface will act as a very good shield at the particular frequency range.

Finally, the contribution of DNG media in the quality of two classical absorber linings is explored. The absorbers, whose geometry is described in Figure 7.9, are the 45° twisted pyramids and the array of alternating wedges. In both of them the last dielectric layer has been replaced with a 7.5-mm-thick metamaterial-loaded tile consisting of rectangular SRRs ($l_r = 3.14$ mm, $r = 0.82$ mm, $l_s = 0.26$ mm, $g = 0.21$ mm, $d = 0.18$ mm). The dimensions of the pyramids are $d_A = 0.34$ m and $d_B = 1.22$ m, whereas those of the wedges are $d_A = 0.20$ m and $d_B = 0.82$ m. The computational analysis is conducted via the hybrid second-/higher-order FDTD algorithm that simulates the volume of a $6.2 \times 4.8 \times 7.4$ m semianechoic chamber with a lattice of $125 \times 106 \times 148$ cells. Figure 7.10 gives the reflectivity of the new absorbers as compared to the usual ones for a relatively large frequency range. The enhancement of the absorbers' competence is, indeed, obvious.

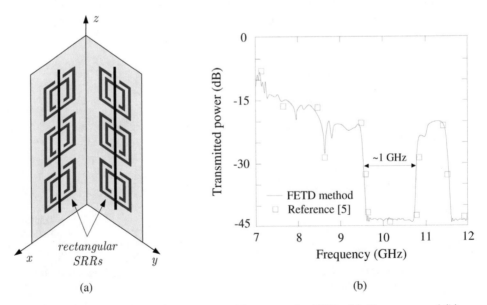

(a)   (b)

FIGURE 7.8: A DNG-based shielding curtain with rectangular SRRs. (a) Geometry and (b) transmitted power.

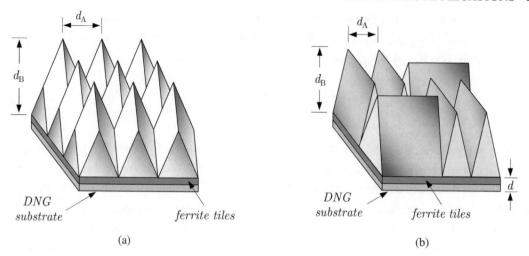

FIGURE 7.9: (a) A DNG-based 45° twisted pyramid absorber lining and (b) an array of alternating DNG-loaded wedges.

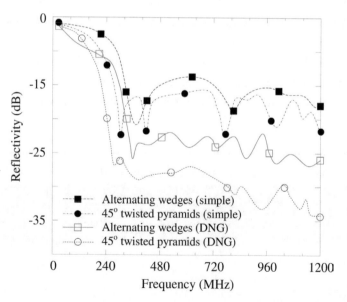

FIGURE 7.10: Reflectivity of alternating wedges and 45° twisted pyramids backed by a DNG substrate.

7.2.4   Performance Optimization of Arbitrarily Sized DNG Slabs

For the study of DNG slabs with either finite or infinite length, the two functions of the magnetic source distribution $\mathbf{K}(x, t) = f(x)g(t)\hat{\mathbf{z}}$ in (7.9) and (7.10), are given by

$$f(x) = e^{-(x/w_0)^2}, \quad g(t) = \begin{cases} g_{on}(t)\sin(2\pi f_0 t), & t < mT_p \\ \sin(2\pi f_0 t), & t \geq mT_p \end{cases} \quad (7.12)$$

with w_0 the pulse width, $g_{on}(t) = 10x_{on}^3 - 15x_{on}^4 + 6x_{on}^5$, $x_{on} = 1 - (mT_p - t)/(mT_p)$, and $T_p = 2\pi/\omega_0$. Moreover after extensive simulations, a range of 9000–10,000 time-steps has been proven fairly sufficient for the steady state to be achieved. At such temporal instants, $E_{tot}(x,y,t) = A(x,y)\sin[2\pi f_0 t + \varphi(x,y)]$, $= \{\tilde{A}(x, y)e^{j2\pi f_0 t}\}$, where $A(x,y)$ is the amplitude, $\varphi(x,y)$ the phase, and $\tilde{A}(x, y) = A(x, y)e^{j\varphi(x, y)}$ the complex envelope at any point of the domain. The last three quantities are calculated in terms of the field magnitude at two different time-steps t_1 and t_2 for which $|t_1 - t_2| < T_p$. Consequently, they are given by

$$A(x,y) = \left|\tilde{A}(x,y)\right| = \left| \frac{E_{tot}(x,y,t_2)e^{-j2\pi f_0 t_1} - E_{tot}(x,y,t_1)e^{-j2\pi f_0 t_2}}{\sin\left[2\pi f_0(t_2 - t_1)\right]} \right| \quad \text{and} \quad \varphi(x,y) = \arg\left[\tilde{A}(x,y)\right] \quad (7.13)$$

7.2.4.1 Infinite and Finite Planar DNG Slabs. Starting with planar metamaterial slabs, spatial increments are selected to be $\Delta x = \Delta y = 0.5$ mm and the temporal one $\Delta t = 1$ ps. The unbounded domain is truncated by a PML absorber, which is appropriately modified to cope with the metamaterial constitutive profiles. In this context, the focusing efficiency of a finite lossy slab is explored and compared with that of its infinite counterpart [30]. Table 7.2 provides the $mse_m(x)$ for both cases, with the finite structure having a length of $\ell = 10\lambda_0$. Undoubtedly, the infinite slab attains a better focusing, especially for low losses and narrow excitation pulses (e.g., $w_0 = 0.1667\lambda_0$). However, as Im$\{n\}$ increases, the finite slab tends *to* exhibit the behavior of the infinite one with the same refrac-

TABLE 7.2: Minimum $mse_m(x)$ of an infinite and a finite DNG Slab with $d = 3.2\lambda_0$ and $\ell = 10\lambda_0$

w_0	$n = -1.0 - j0.001$		$n = -1.0 - j0.1$		$n = -0.9 - j0.001$		$n = -1.1 - j0.001$	
	INFINITE	FINITE	INFINITE	FINITE	INFINITE	FINITE	INFINITE	FINITE
$0.1667\lambda_0$	0.2777	0.3047	0.8772	0.8806	0.39	0.4056	0.3523	0.3682
$0.3333\lambda_0$	0.0304	0.0403	0.8049	0.8098	0.0908	0.0998	0.0674	0.0712
$0.5\lambda_0$	0.0016	0.0024	0.7751	0.7804	0.0136	0.0179	0.0073	0.0086

tive index. These remarks state that, for low losses, the finite length—unlike its infinite analogue—deteriorates the slab's performance. In contrast, for high Im$\{n\}$, the lossy profile compensates the influence of the finite length.

An alternative way to estimate the focusing properties of a finite slab may be pursued from the transmission coefficient $T(k_y)$ between the source and the focus plane. Figure 7.11 presents its amplitude and phase for various n. As observed, the slab cannot sufficiently reconstruct all wave modes, even for the lossless case. Particularly, for propagating modes ($k_y < k_0$), when $\ell = 10\lambda_0$ and $d = 3.2\lambda_0$, the transmission coefficient amplitude is close to its maximum apart from some small fluctuations (Figure 7.11a). However, as $k_y > k_0$ $k_y > k_0$ (evanescent wave region), $|T|$ decreases very rapidly.

Concerning the phase in Figure 7.11b, it remains nearly zero up to a specific wavenumber near the propagating/evanescent boundary. It can also be extracted that, as losses augment, the plots coincide with those of an infinite slab. This occurs because losses do not affect all wave modes to the same degree. In fact, wave modes with large k_y accept a greater attenuation than those with small k_y. Regarding the infinite slab, the presence of nonzero Im$\{n\}$ leads to the complete annihilation of a significant portion of evanescent wave modes at the focus plane [30]. Thus, only propagating waves with reduced amplitude reach the focus plane. In contrast, in the finite case, evanescent waves on the focus plane vanish, even when Im$\{n\} = 0$. Thus, losses attenuate exclusively the amplitude of propagating waves in exactly the same manner as they do at the infinite slab. In conclusion, at the focus plane, there solely exist propagating modes for both the infinite and finite slabs, a fact implying that both cases exhibit similar behavior. For this phenomenon to arise, Im$\{n\}$ depends on d. Hence, given a specific d, one can determine the minimum value of Im$\{n\}$, above which the two cases become equivalent.

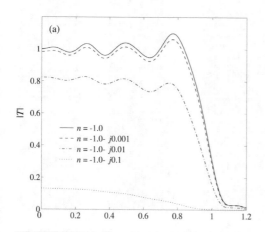
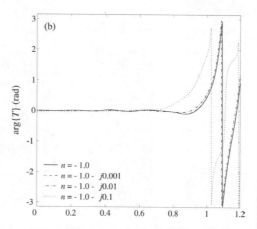

FIGURE 7.11: Computation of the transmission coefficient of a finite lossy DNG slab as function of k_y/k_0 with $\ell = 10\lambda_0$ and $d = 3.2\lambda_0$. (a) Magnitude and (b) phase.

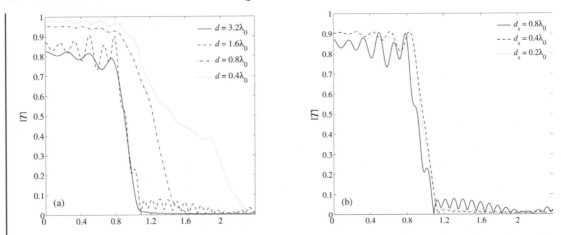

FIGURE 7.12: Transmission coefficient amplitude versus k_y/k_0 of a finite lossy DNG slab with $n = -10 - j0.01$ for (a) $d_s = d/2$, d variable and (b) $d = 1.6\lambda_0$, d_s variable.

Our numerical verification continues with the outcomes of Figure 7.12 that show $|T(k_y)|$ for $n = -10 - j0.01$ and diverse d, d_s. More specifically, in Figure 7.12a, a set of slabs with various widths and $d_s = d/2$ is examined. One may easily deduce that, as d decreases, more wave components with larger k_y are regenerated at the focus plane. For $d = 3.2\lambda_0$, the transmission coefficient approaches zero as soon as reaches the evanescent-wave region. A similar behavior is encountered in the $d = 1.6\lambda_0$ case as well, but with more fluctuations.

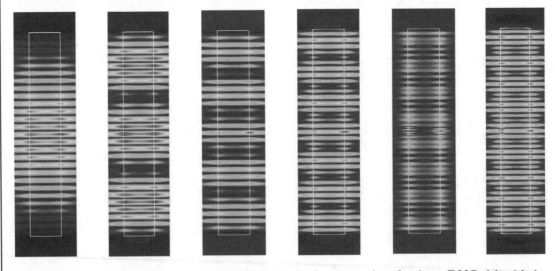

FIGURE 7.13: Spatial profile of the electric field for the first six modes of a planar DNG slab with $\ell = 10\lambda_0$ and $d = 3.2\lambda_0$.

Nonetheless, this situation is modified for $d = 0.8\lambda_0$ and $d = 0.4\lambda_0$, where a large part of the evanescent-wave modes crosses the slab. To determine the reason of this change, whether it is the smaller width or the closer distance from the source, the transfer function for a slab of fixed d and a varying d_s is presented in Figure 7.12b. Despite a visible improvement of the variations magnitude, no other notable modification of the cutoff wavenumber can be detected. Such an issue means that the principle way for enhancing the slab's focusing aptitude is its width, regardless of source position.

Lastly, Figure 7.13 illustrates the spatial profile of field amplitude for the first six modes (the dominant one included), while $\ell = 10\lambda_0$ and $d = 1.6\lambda_0$. In all cases, the field is confined in $-\ell / 2 \le y \le \ell / 2$ span, thus proving the claim that the finite length leads to a resonator-like performance.

7.2.4.2 Adjustable-Angle Triangular Lossy DNG Slabs.
Moving from planar slabs to wedge-like configurations, assume the DNG-metamaterial right triangle of Figure 7.14a with $\varepsilon_r = \mu_r = -1.0 - j\delta$ and refractive index n. The lengths of its normal sides AC and BC are l_x and l_y, whereas the angle between AC and the hypotenuse AB is θ_0. The structure is illuminated by an electromagnetic source set on a parallel to AC plane at distance d_s. Neglecting the lossy term δ and assuming a point source at S, in order its projection on AC to lie at distance l_s from A, it can be shown that the overall energy focuses on two points F_1 and F_2, in the interior and exterior of the triangle, respectively. Actually, there are two ways to investigate such structures: via a ray approximation method for the field or through the Poynting vector dynamic lines.

According to the former, Snell's law at a vacuum-DNG interface gives $\theta_t = -\theta_i$, where θ_i is the incidence angle and θ_t is the refractive one. Thus, an arbitrary ray, emanating from S and forming an angle θ_{ri} with AC, is bent toward the opposite direction and at the same angle, $\theta_{rt} = \theta_{ri}$, in the meta-

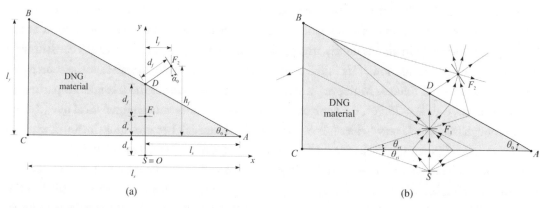

FIGURE 7.14: (a) Geometry of a DNG right triangle illuminated by a point source at S and (b) ray approximation of the field generated at S as it crosses the triangle.

material, as depicted in Figure 7.14b. If the same procedure is repeated for all rays, it is derived that they intersect at F_1, allowing for the electromagnetic energy, generated at S, to focus at F_1. Similarly, one may prove the existence of another focus point F_2 outside the triangle on the right-hand side of AB. However, the rays forming an angle greater than $\pi - \theta_0$ with AC, cannot "reach" AB and cross F_2, despite the fact that they do cross F_1. Hence, the field at F_2 is *not* expected to be a perfect image of its counterpart at the source plane. The positions of points F_1 and F_2 are obtained by d_f, l_f, h_f lengths, expressed in terms of d_s, l_s, θ_0, as

$$d_f = l_s \tan \theta_0 - d_s, \qquad (7.14)$$

$$l_f = 2(l_s \sin \theta_0 - d_s \cos \theta_0) \sin \theta_0, \qquad (7.15)$$

$$h_f = d_s + 2(l_s \sin \theta_0 - d_s \cos \theta_0) \cos \theta_0 \qquad (7.16)$$

Although the prior formulation involves a point source at S, a distributed one on the source plane is generally used. For this goal, an orthogonal coordinate system with its origin at S, x-axis parallel to AC, and y-axis parallel to SF_1, is defined. Also, the incident field is assumed to be TE, i.e., only the E_z component is nonzero. Again, the distributed source is a magnetic current sheet $\mathbf{K}(x, t) = f(x)g(t)\hat{\mathbf{z}}$ on the $y = 0$ plane, with $f(x)$ and $g(t)$ mathematically described as in the planar DNG slab case. Essentially, the key issue for such structures is to determine the quality of field reconstruction at the focus planes. Because only the external to the triangle focus point is of practical importance, the analysis will mainly concentrate on it.

The first investigation concerns the impact of the adjustable angle θ_0 on the behavior of electric field amplitude. Three different values for a right triangle are examined with $l_x = 600\Delta x$. The structure is illuminated by a source of $l_x = 300\Delta x$, $d_s = 50\Delta y$, and pulse width $w_0 = 20\Delta x$, whereas the unbounded space is terminated by an eight-cell PML. Figure 7.15 provides the corresponding surface plots where the dashed lines denote the beam axes and their small perpendicular ones the respective source and focus planes. Evidently, one can discern that electromagnetic energy, generated at the source plane, focuses initially inside the triangle [30]. In fact, the reconstructed field constitutes a very satisfactory representation of its initial counterpart, because the beam waists at the source and focus planes are almost equal. Moreover, a second focus occurs outside the triangle. In particular, for small angles, e.g., $\theta_0 = 25.56°$, the computed beam axis coincides with the analytical one (dashed line), although the beam is now "wider." As θ_0 increases, the beam axis becomes more inclined, whereas the "beam center" (point where electric field amplitude is maximized) still remains at F_2.

Similar deductions are drawn via the Poynting vector dynamic lines of Figure 7.16. Therefore, for the triangle with the smallest θ_0 angle, its energy flow, at the focus plane is parallel to the theoretical beam axis. Nevertheless, further increase of θ_0 drives the dynamic lines to greater inclinations toward this axis. This is mainly accredited to the fact that any beam can be considered a sum

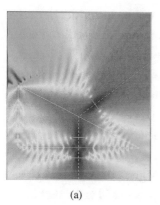
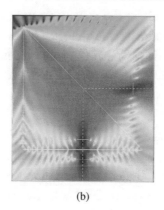

| (a) | (b) | (c) |

FIGURE 7.15: Surface plots of the electric field amplitude at the steady state, for three different right triangles: (a) $\theta_0 = 25.56°$, (b) $\theta_0 = 45°$, and (c) $\theta_0 = 63.43°$. The dashed lines indicate the theoretically predicted beam axes with their small perpendicular ones denoting the respective source and focus positions.

of plane waves, with each dynamic line (adequately distanced from the beam waist) determining the propagation direction of a single plane-wave component. Thus, if such waves travel symmetrically along an arbitrary line—or if for any quantity whose propagation vector forms an angle φ_i with the line, there exists its symmetric equivalent of the same amplitude and angle $-\varphi_i$—then the beam axis will coincide with this line. On the contrary, if there exist plane waves without any symmetric analogues, the beam axis appears inclined to the prior line. This is exactly the case for the triangular lossy DNG slabs with large θ_0 values. Specifically, some plane-wave components that propagate on the left of the theoretical beam axis in the metamaterial deviate to the triangle's left side and do not

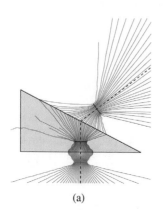
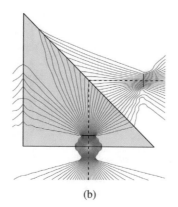
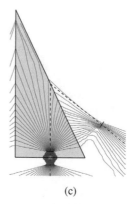

| (a) | (b) | (c) |

FIGURE 7.16: Poynting vector dynamic lines for three different right triangles: (a) $\theta_0 = 25.56°$, (b) $\theta_0 = 45°$, and (c) $\theta_0 = 63.43°$. The dashed lines indicate the theoretically predicted beam axes with their small perpendicular ones denoting the respective source and focus positions.

"reach" the focus plane. As a result, at the external focus point, there will be quantities devoid of symmetric constituents, hence enforcing the evaluated beam to incline toward the predicted one.

Next, the benefits of criteria (7.6) and (7.7) are validated for a triangular DNG slab of $\theta_0 =$ 45°. Returning to Figure 7.14a, it can be derived that, except the source plane, the beam waist arises at a parallel to AC plane, which contains F_1, and at an α_0-inclined one along y-axis, where F_2 belongs to.

Owing to the confined size of the computational domain, the infinite integration in the two relations is approximated by a finite one in a bounded space $[u_{min}, u_{max}]$ on the condition that externally the field tends to zero. Under these assertions, Figure 7.17a gives the cross-correlation coefficient for various inclination angles α. Obviously, the plane on which the field is the best image of its counterpart at the source plane (i.e., where $\rho(s)$ receives its maximum value) does not concur with the theoretical one as already detected from the outcomes of Figures 7.15 and 7.16. Conversely, the mean-square error, with and without the extra exponential term in (7.7), is illustrated in Figure 7.17b. Again, it is discerned that the enhanced $mse_m(s)$ outperforms the usual $mse(s)$ criterion, because of its completely smooth shape.

Finally, Table 7.3 provides the α^{num} values for the optimal observation plane along with the maximum $\rho(s)$ and minimum $mse_m(s)$ for triangles with various θ_0 and refractive index imaginary parts (losses). From the results, the focusing efficiency of the DNG triangle deteriorates or equivalently $\rho(s)$ decreases and $mse_m(s)$ increases. Nonetheless, when $Im\{n\} = 0.01$, the maximum $\rho(s)$ seems to augment somehow, with the minimum $mse_m(s)$ rising according to our expectations. A vi-

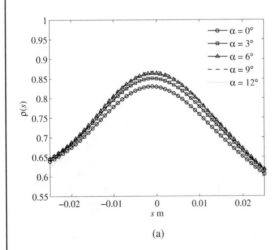
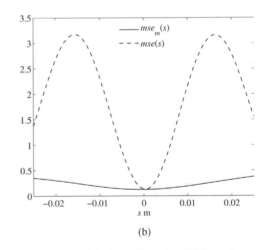

(a)   (b)

FIGURE 7.17: (a) Cross-correlation coefficient versus s for a triangle with $\theta_0 = 45°$ and variable inclination angle, α, of the observation plane. (b) Original and enhanced mean-square error as a function of s for $\theta_0 = 45$ and $\alpha = 40°$.

TABLE 7.3: Numerically estimated inclination angle of the optimum observation plane, maximum $\rho(s)$ and minimum $\mathrm{mse}_m(s)$ for various θ_0 and refractive indices

θ_0	26.56°	33.69°	39.80°	45°	53.13°	59.03°	63.43°
$n = -1.0 - j0.01$							
α_0^{num}	40°	29°	19°	11°	−1°	−11°	−17°
$\max\{\rho(s)\}$	0.9350	0.9132	0.8917	0.8871	0.8486	0.8314	0.8192
$\min\{\mathrm{mse}_m(s)\}$	0.1009	0.1451	0.1880	0.2094	0.2994	0.3503	0.3858
$n = -1.0 - j0.01$							
α_0^{num}	45°	35°	27°	21°	11°	3°	−3°
$\max\{\rho(s)\}$	0.8771	0.8499	0.8725	0.8136	0.7891	0.7693	0.7447
$\min\{\mathrm{mse}_m(s)\}$	0.7083	0.8205	0.8835	0.9189	0.9572	0.9726	0.9802

able explanation is furnished if one observes that $\mathrm{mse}_m(s)$ is a practically stricter criterion than $\rho(s)$. Evenly, although the field at the observation plane exhibits a set of rather dissimilar amplitudes for refractive indices $n = -10 - j0.01$ and $n = -10 - j0.1$, its shape is only slightly divergent.

7.3 BI-ISOTROPIC AND CHIRAL MEDIA

Bi-isotropic media are a general family of materials that exhibit magnetoelectric coupling such that an arbitrary excitation concurrently produces both electric and magnetic polarization [32–43]. On the other hand, chiral media probably constitute the most promising subcategory of bi-isotopic structures. Their application record is indeed impressive, ranging from waveguide mode converters and radar absorbers to EMI shielding and impedance transformers. Of critical importance in the study of a chiral medium is its degree of polarization coupling or its "chirality parameter," which has motivated an intensive research. Toward this direction, additional momentum has been imposed by the recent investigations on the properties of DNG metamaterials.

7.3.1 Dispersive FDTD Schemes for Complex Wave Interactions
The basic difference between bi-isotropic and conventional dielectric or magnetic media is located in the magnetoelectric coupling that introduces extra terms to the usual constitutive relations. Spe-

cifically, these expressions for a general bi-isotropic medium may be conveniently written in the subsequent form of

$$\mathbf{D} = \varepsilon\mathbf{E} + \xi\mathbf{H}, \quad \mathbf{B} = \zeta\mathbf{E} + \mu\mathbf{H}, \qquad (7.17)$$

where

$$\xi = (\chi - j\kappa)(\mu_0\varepsilon_0)^{1/2}, \quad \zeta = (\chi + j\kappa)(\mu_0\varepsilon_0)^{1/2}, \qquad (7.18)$$

where χ is the Tellegen and κ is the chirality parameter. Working in the frequency domain and assuming the existence of the $e^{j\omega t}$ convention, Maxwell's equations, via (7.17), become

$$\nabla \times \mathbf{E}(\omega) = -j\omega[\mu\mathbf{H}(\omega) + \zeta\mathbf{E}(\omega)], \quad \nabla \times \mathbf{H}(\omega) = j\omega[\varepsilon\mathbf{E}(\omega) + \xi\mathbf{H}(\omega)], \qquad (7.19)$$

To avoid the unnecessary introduction of supplementary forms in the time-domain analogues of (7.19) and therefore extract workable explicit update formulae, electric and magnetic fields inside the bi-isotropic medium are decomposed as $\mathbf{E} = \mathbf{E}_+ + \mathbf{E}_-$ and $\mathbf{H} = \mathbf{H}_+ + \mathbf{H}_-$. These vector quantities can be deemed circularly polarized waves, where the "+" indicates the right-hand and "−" the left-hand circular polarization [35]. Actually, $\mathbf{E}_\pm$ and $\mathbf{H}_\pm$ wave fronts consider the bi-isotropic medium as an equivalent isotropic substance with effective constitutive parameters ε_+, ε_-, μ_+, and μ_-, given by

$$\varepsilon_\pm = \varepsilon(\cos\theta \pm \kappa_r)e^{\pm j\theta}, \quad \mu_\pm = \varepsilon(\cos\theta \pm \kappa_r)e^{\pm j\theta}, \qquad (7.20)$$

where $\theta = \sin^{-1}(\chi_r)$ and κ_r, χ_r are the chirality and Tellegen parameters, respectively, normalized by the refractive index n_r. Note that κ_r is a measure of the material's nonreciprocity and when $\kappa_r = 0$, it represents a reciprocal chiral medium. Moreover, because the two sets of fields are independent and do not couple in a homogeneous bi-isotropic medium, due to the preceding decomposition, (7.19) can be written as

$$\nabla \times \mathbf{E}_\pm(\omega) = -j\omega\mu_\pm\mathbf{H}_\pm(\omega), \quad \nabla \times \mathbf{H}_\pm(\omega) = j\omega\varepsilon_\pm\mathbf{E}_\pm(\omega) \qquad (7.21)$$

The main advantage of this approach is that (7.21) in the equivalent media can be easily discretized and incorporated into a standard time-domain formulation, like the FDTD method, for the case where the material parameters are not dispersive. The resulting scheme is fully explicit and consequently does not increase the total memory requirements when compared to conventional constituents. Thus, the second of (7.21) gives

$$\nabla \times \mathbf{H}_\pm(\omega) = j\omega\left(\varepsilon \pm j\frac{\xi_r}{\eta}e^{\pm j\theta}\right)\mathbf{E}_\pm(\omega), \qquad (7.22)$$

for $\xi_r = (\chi_r - j\kappa_r)(\mu_0\varepsilon_0)^{1/2}$ derived from (7.18). Focusing on a reciprocal medium ($\chi_r = 0$), it is obtained that $\xi_r = -j\kappa_r(\mu_0\varepsilon_0)^{1/2}$. Then, (7.21) can be rewritten in the time domain as

$$\nabla \times \mathbf{H}_{\pm} = \varepsilon(1 \pm \kappa_r) \frac{\partial}{\partial t} \mathbf{E}_{\pm}, \qquad (7.23)$$

which, for the 2-D lossless and source-free chiral TE case, leads to the following update relations

$$E_{z\pm}|_{i-1/2,j+1/2}^{n+1} = E_{z\pm}|_{i-1/2,j+1/2}^{n} + \frac{\Delta t}{\varepsilon_0 \varepsilon_r (1 \pm \kappa_r)}$$

$$\times \left(\frac{H_{y\pm}|_{i,j+1/2}^{n+1/2} - H_{y\pm}|_{i-1,j+1/2}^{n+1/2}}{\Delta x} + \frac{H_{x\pm}|_{i-1/2,j}^{n+1/2} - H_{x\pm}|_{i-1/2,j+1}^{n+1/2}}{\Delta y} \right), \qquad (7.24)$$

$$H_{x\pm}|_{i-1/2,j+1}^{n+1/2} = H_{x\pm}|_{i-1/2,j+1}^{n-1/2} + \frac{\Delta t}{\mu_0 \mu_r (1 \pm \kappa_r) \Delta y} \left(E_{z\pm}|_{i-1/2,j+1/2}^{n} - E_{z\pm}|_{i-1,j+3/2}^{n} \right), \qquad (7.25)$$

$$H_{y\pm}|_{i,j-1/2}^{n+1/2} = H_{y\pm}|_{i,j-1/2}^{n-1/2} + \frac{\Delta t}{\mu_0 \mu_r (1 \pm \kappa_r) \Delta x} \left(E_{z\pm}|_{i+1/2,j+1/2}^{n} - E_{z\pm}|_{i-1/2,j+1/2}^{n} \right) \qquad (7.26)$$

Observe that the above stencil selections are not unique, because several forward- and backward-differencing schemes may be used. On the other hand, for the frequency-dependent case, it holds

$$\mathbf{D}_{\pm}(\omega) = [\varepsilon(\omega) \pm \varepsilon_0 \kappa_r(\omega)] \mathbf{E}_{\pm}(\omega), \quad \mathbf{B}_{\pm}(\omega) = [\mu(\omega) \pm = \mu_0 \kappa_r(\omega)] \mathbf{H}_{\pm}(\omega) \qquad (7.27)$$

Next, a Lorentz model for the permittivity and permeability, and a Condon model for the chirality parameter to approximate the mechanism of the dispersive material, are used, namely,

$$\varepsilon(\omega) = \varepsilon_0 \left[\varepsilon_\infty + \frac{(\varepsilon_s - \varepsilon_\infty)\omega_e^2}{\omega_e^2 + j2\Gamma_e\omega - \omega^2} \right], \quad \mu(\omega) = \mu_0 \left[\mu_\infty + \frac{(\mu_s - \mu_\infty)\omega_m^2}{\omega_m^2 + j2\Gamma_m\omega - \omega^2} \right], \qquad (7.28)$$

$$\kappa(\omega) = \frac{\omega\omega_c}{\omega_c^2 + j\Gamma_c\omega\omega_c - \omega^2}, \qquad (7.29)$$

where ω_e, ω_m, and ω_c are the electric resonance, the magnetic resonance, and the magnetoelectric coupling frequencies, respectively, while Γ_e, Γ_m, and Γ_c denote the corresponding damping factors [39]. In this framework, the time-domain version of (7.27), presuming a causal frequency dependence, becomes

$$\mathbf{D}_{\pm} = \varepsilon_0 \varepsilon_\infty \mathbf{E}_{\pm} + \varepsilon_0 \int_0^t \mathbf{E}_{\pm}(t - \tau) [\chi_e(\tau) \pm \kappa_r(\tau)] d\tau, \qquad (7.30)$$

$$\mathbf{B}_{\pm} = \mu_0 \mu_\infty \mathbf{H}_{\pm} + \mu_0 \int_0^t \mathbf{H}_{\pm}(t - \tau) [\chi_m(\tau) \pm \kappa_r(\tau)] d\tau, \qquad (7.31)$$

with $f = \chi_e, \chi_m, \kappa_r$ having the form of $f = \text{Re}\{-j\gamma_p e^{-(\alpha_p + j\beta_p)t}\}u(t)$, where index p = e, m, c refers to each of the f parameters and $u(t)$ the unit-step function. In particular,

$$\alpha_e = \Gamma_e, \qquad \beta_e = \sqrt{\omega_e^2 - \alpha_e^2}, \qquad \gamma_e = (\varepsilon_s - \varepsilon_\infty)\beta_e^{-1}\omega_e^2$$

$$\alpha_m = \Gamma_m, \qquad \beta_m = \sqrt{\omega_m^2 - \alpha_m^2}, \qquad \gamma_m = (\mu_s - \mu_\infty)\beta_m^{-1}\omega_m^2$$

$$\alpha_c = \Gamma_c/2, \qquad \beta_c = \sqrt{\omega_c^2 - \alpha_c^2}, \qquad \gamma_c = \sqrt{\omega_c^2 + \rho_c^2}, \qquad \rho_c = -\omega_c^2 a_c \beta_c^{-1}$$

Discretizing the continuous temporal variable t as $n\Delta t$ in (7.30), (7.31) and applying the concept of piecewise recursive convolution, the electric flux density may be expressed as

$$\mathbf{D}_\pm^n = \varepsilon_0 \varepsilon_\infty \mathbf{E}_\pm^n + \varepsilon_0 \sum_{\ell=0}^{n-1} \mathbf{E}_\pm^{n-\ell}\left(\chi_e^\ell \pm \kappa_{r,e}^\ell\right) + \left(\mathbf{E}_\pm^{n-\ell-1} - \mathbf{E}_\pm^{n-\ell}\right)\left(\xi_e^\ell \pm \zeta_e^\ell\right), \qquad (7.32)$$

with a similar formula holding for the magnetic flux density. In (7.32),

$$\chi_e^\ell = \int_{\ell\Delta t}^{(\ell+1)\Delta t} \chi_e(\tau)\mathrm{d}\tau, \qquad \kappa_e^\ell = \int_{\ell\Delta t}^{(\ell+1)\Delta t} \kappa_r(\tau)\mathrm{d}\tau,$$

$$\xi_e^\ell = \int_{\ell\Delta t}^{(\ell+1)\Delta t} (\tau - \ell\Delta t)\chi_e(\tau)\mathrm{d}\tau, \qquad \zeta_e^\ell = \int_{\ell\Delta t}^{(\ell+1)\Delta t} (\tau - \ell\Delta t)\kappa_r(\tau)\mathrm{d}\tau$$

Basically, these relations comprise the source for the extraction of the final update schemes

$$\mathbf{E}_\pm^{n+1} = \frac{1}{\Theta_e^0}\left[\left(\varepsilon_\infty - \xi_e^0 \mp \zeta_e^0\right)\mathbf{E}_\pm^n + \frac{\Delta t}{\varepsilon_0}\nabla \times \mathbf{H}_\pm^{n+1/2} + \mathbf{R}_{e\pm}^n \pm \mathbf{R}_{c\pm}^n\right], \qquad (7.33)$$

$$\mathbf{H}_\pm^{n+1/2} = \frac{1}{\Theta_m^0}\left[\left(\mu_\infty - \xi_m^0 \mp \zeta_m^0\right)\mathbf{E}_\pm^n - \frac{\Delta t}{\mu_0}\nabla \times \mathbf{E}_\pm^{n+1} + \mathbf{Q}_{m\pm}^{n-1/2} \pm \mathbf{Q}_{c\pm}^{n-1/2}\right], \qquad (7.34)$$

where $\Theta_e^0 = \varepsilon_\infty - \xi_e^0 + \chi_e^0 \pm \kappa_{r,e}^0 \mp \zeta_e^0$ and $\Theta_m^0 = \mu_\infty - \xi_m^0 + \chi_m^0 \pm \kappa_{r,m}^0 \mp \zeta_m^0$. The other vectors are obtained via the sums in (7.32) and the possibility of expressing the chirality parameter as an exponential. For example,

$$\mathbf{R}_{e\pm}^n = \mathrm{Re}\left\{\mathbf{E}_\pm^n \Delta \kappa_{r,e}^0 \pm \mathrm{e}^{(-\alpha_c + j\beta_c)\Delta t}\sum_{\ell=1}^{n-1} \mathbf{E}_\pm^{n-\ell}\Delta \kappa_{r,e}^\ell\right\}, \qquad (7.35)$$

$$\mathbf{Q}_{m\pm}^{n-1/2} = \mathrm{Re}\left\{\mathbf{H}_\pm^{n-1/2}\Delta \kappa_{r,m}^0 \pm \mathrm{e}^{(-\alpha_c + j\beta_c)\Delta t}\sum_{\ell=1}^{n-1} \mathbf{H}_\pm^{n-\ell-1/2}\Delta \kappa_{r,m}^\ell\right\}, \qquad (7.36)$$

for

$$\Delta \kappa_{r,e}^\ell = \kappa_{r,e}^\ell - \kappa_{r,e}^{\ell+1} = \frac{j\gamma_c}{(-\alpha_c + j\beta_c)}\mathrm{e}^{(-\alpha_c + j\beta_c)\ell\Delta t}\left[1 - \mathrm{e}^{(-\alpha_c + j\beta_c)\Delta t}\right]^2,$$

$$\Delta \kappa_{r,m}^\ell = \kappa_{r,m}^{\ell-1/2} - \kappa_{r,m}^{\ell+1/2} = \frac{j\gamma_c}{(-\alpha_c + j\beta_c)}\mathrm{e}^{(-\alpha_c + j\beta_c)(\ell+1/2)\Delta t}\left[1 - \mathrm{e}^{(-\alpha_c + j\beta_c)\Delta t}\right]^2,$$

and

$$\kappa_{r,e}^{0} = \mathrm{Re}\left\{\frac{j\gamma_c}{(-\alpha_c + j\beta_c)}\left[1 - e^{(-\alpha_c + j\beta_c)\Delta t}\right]\right\},$$

$$\kappa_{r,m}^{0} = \mathrm{Re}\left\{\frac{j\gamma_c}{(-\alpha_c + j\beta_c)}e^{(-\alpha_c + j\beta_c)\Delta t/2}\left[1 - e^{(-\alpha_c + j\beta_c)\Delta t}\right]\right\}$$

In this way, all electric and magnetic field components can evolve in a recurrence regime and the most significant: their computation hardly increases the system overhead due to the absence of auxiliary equations.

7.3.2   Numerical Simulations

The first simple, yet indicative, application deals with the propagation of a modulated Gaussian pulse in a 1-D bi-isotropic medium. For this purpose, the computational domain is divided into

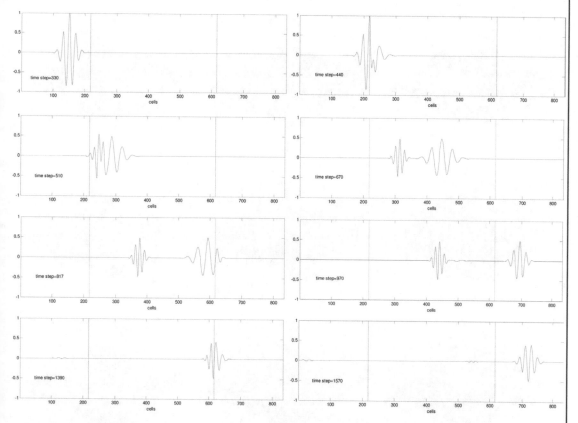

FIGURE 7.18: Temporal evolution of an E_x modulated Gaussian-pulse component in a bi-isotropic medium (accommodating the area between the lines) for $\Delta t = 0.4\Delta/v$, $\kappa_r = 0.6$, $T = 40\Delta t$, and $\lambda = 20\Delta$.

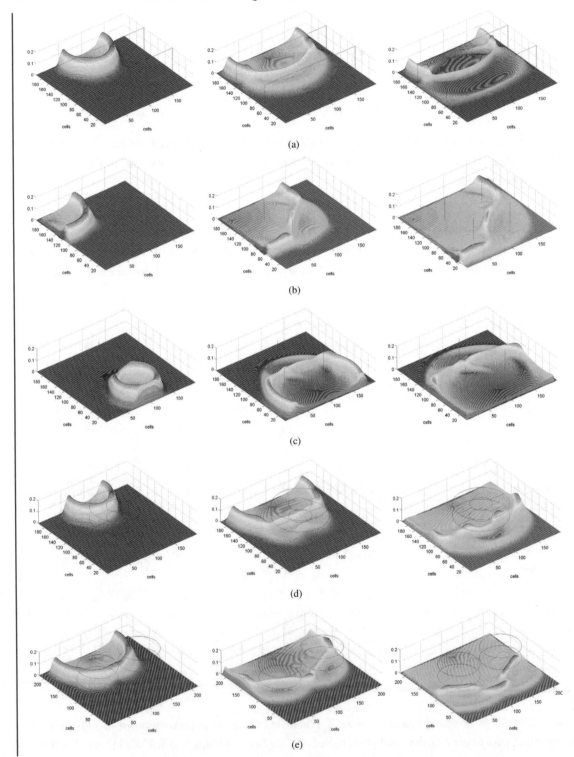

three regions, with the second one containing the bi-isotropic slab and the other two filled with air. The open boundaries are backed by an eight-cell PML, whereas the excitation is launched as a hard source. Figure 7.18 presents a set of characteristic snapshots of the E_x component for $\Delta t = 0.4\Delta/\upsilon$ (Δ is the spatial increment and υ the phase velocity), $\kappa_r = 0.6$, $T = 40\Delta t$, and $\lambda = 20\Delta$. It becomes apparent that as soon as the pulse enters the bi-isotropic area, it is separated into two distinct modes, namely, a left-handed and a right-handed one, so fully confirming the theoretical analysis of the previous paragraph.

Essentially, the former mode seems to have the larger phase velocity, whereas the latter is always the slowest. To this end, the duration of the left-handed mode augments with its entrance to the material (spatial dispersion), in contrast to the one of its right-hand counterpart. In addition, one may possibly come up with the conclusion that the pulse duration T increases with the rise of κ_r (reduction of Δt) and the most important: the separation of the incident wavefield becomes *more abrupt* as κ_r gets larger values, namely, the difference between the modes' phase velocity becomes more discernible. Lastly, during the entrance and the exit of the pulses from the bi-isotropic slab some small reflections are observed, possibly attributed to the inherent traits of the FDTD method [43].

Proceeding to 2-D applications, Figure 7.19 presents a set of H_z snapshots for various bi-isotropic scatterers that may be used as intermediate layers in the construction of efficient high-frequency absorbers for EMC test facilities. In fact, these linings are capable of dispersing the electromagnetic energy to supplementary directions, hence relieving the quiet zone of the chamber from unwanted parasites. Specifically, a rectangular, a square, a triangular, and a toroidal case with diverse T and κ_r values are examined.

Besides, Figure 7.19e explores the propagation of the H_z component for a couple of cylindrical bi-isotropic scatterers. From the results, the separation of the pulse in the interior of all scatterers is clearly visible without any instabilities or accuracy discrepancies. Also, the PML truncation seems to successfully cope with the majority of the relatively involved modes that leave the material and does not produce any artificial oscillations. Therefore, it can be safely deduced that the utilization of the dispersive FDTD algorithm, described in Section 7.3.1 can constitute a very reliable choice for the time-domain analysis of such media.

7.4   NANOSTRUCTURES AND NANOTECHNOLOGY APPLICATIONS

Since the first fabrication of EMC nanostructures, extensive research and experimentation in radiating patches and arrays has led to structures with different operational characteristics that exhibit

---

FIGURE 7.19: Snapshots of the H_z component through (a) a rectangular slab ($T = 20\Delta t$, $\kappa_r = 0.4$), (b) a square scatterer ($T = 40\Delta t$, $\kappa_r = 0.2$), (c) a triangular scatterer ($T = 30\Delta t$, $\kappa_r = 0.5$), (d) a toroidal scatterer ($T = 20\Delta t$, $\kappa_r = 0.35$), and (e) two cylindrical scatterers ($T = 40\Delta t$, $\kappa_r = 0.4$).

numerous advantages [44–48]. In recent years, these types of microstrip antennas have become more complicated due to the inhomogeneous and anisotropic features of their dielectric substrates, the inclusion of loads between the patch and the ground plane, and the integration of lumped circuitry within the antenna body. Of critical significance as well is the polarization control of the radiator, which due to the miniaturized dimensions requires a very careful handling. In particular, circular polarization operations can be obtained by embedding slots or inserting slits of different length at the edges or truncating the patch corners. Furthermore, the use of loading slots in the patch may accomplish a robust dual-frequency operation, where the resonant frequencies are tuned via the length of the slots. Such patches, nevertheless, have demanding shapes and substrates that render their implementation laborious or even practically unaffordable, at least through the "trial and error" logic.

Here, two compact single-feed corner-truncated nanostructures (Figure 7.20) with dispersive substrates, not previously analyzed by a numerical technique, are investigated [49, 50]. The return loss and the input impedance in a wide-frequency band are extracted and the influence of dielectric dispersion is also examined. All simulations are conducted by means of the frequency-dependent FDTD approach, although different selections are also possible. It is emphasized that the dielectric properties of the substrate depend on the excitation frequency, the operation temperature, the water absorption, and even the ultraviolet radiation exposure. Evidently, the incorporation of all these fac-

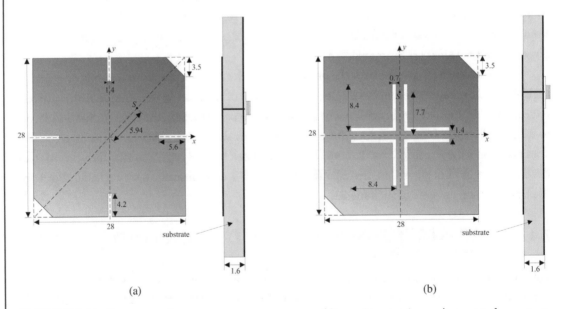

(a)   (b)

FIGURE 7.20: Geometry of two nanostructure antennas (dimensions are in mm) mounted on a rectangular patch with two truncated corners. Configuration (a) with two pairs of narrow edge slits and (b) with four L-shaped central slits. Letter S indicates the feeding point of the patch.

tors in a single computational model is very difficult, yet the effect of material dispersion on the antenna operation is often attainable. Usually, material manufacturers specify the electric loss tangent tan δ at a set of frequencies. For instance, for RT/Duroid 5880 a loss tangent of tan $\delta = 9 \times 10^{-4}$ at 10 GHz and a relative dielectric constant of 2.2 are usually provided. Because in nature, the relative permittivity and the loss tangent are functions of the excitation frequency, a more precise description of their electric behavior is through the well-known Debye models. In the following examples, an one-pole model for the rigorous representation of an epoxy substrate is presumed that leads to a loss tangent of 0.01834 and a real part of the relative permittivity equal to 4.4.

Focusing on the antennas of Figure 7.20, special attention must be paid to the pairs of narrow slits—either edge (Figure 7.20a) or L-shaped central (Figure 7.20b)—and the two truncated corners that control the beam directivity and the gain of the structure. The simulation region is divided into $60 \times 20 \times 20$ cells with a size of $\Delta x = \Delta y = 0.07$ cm and $\Delta z = 0.04$ cm.

The return loss for both arrangements is depicted in Figures 7.21a and 7.22a, respectively. Notice that the resonant frequencies of the lossless model are slightly lower than those of the Debye one, a divergence that becomes greater at higher-order modes. This is basically because the value of the relative dielectric permittivity is lower at these frequencies and the resonant frequency of the patch is inversely proportional to the square root of its substrate relative permittivity. In this context, Figures 7.21b and 7.22b illustrate the input resistance for the first mode. The anticipated differences between the two calculations get more prominent at higher-order modes, so indicating the sensitivity of the antenna at substrate dispersive losses. Actually, such a discrepancy renders the

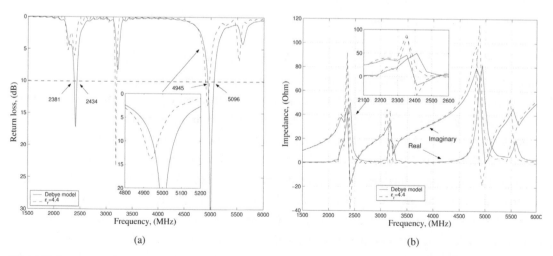

FIGURE 7.21: Computed (a) return loss and (b) input impedance (real and imaginary part) versus frequency for the antenna of Figure 7.20a (Debye model and lossless case).

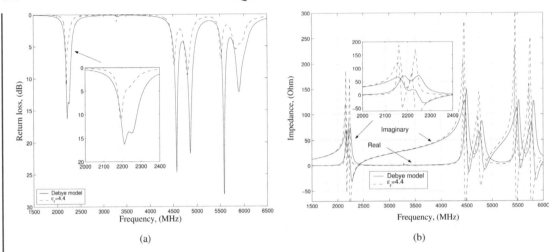

FIGURE 7.22: Computed (a) return loss and (b) input impedance (real and imaginary part) versus frequency for the antenna of Figure 7.20b (Debye model and lossless case).

matching with a coax cable or a microstrip line a tricky procedure. From numerical simulations, it is also discerned that the radiation patterns are not greatly modified and the primary characteristics remain unchanged but at some dB lower in magnitude owing to the dielectric losses of the substrate. As a last remark, the operation bandwidth of these devices is found to be rather small (usually up to 150 MHz) with the dielectric properties of the substrate material being practically constant in this frequency region. However, in modern dual-band antennas, the two operating modes are quite far from each other in the spectrum, with their distance reaching up to 2.5 GHz.

7.5   PERIODIC PHOTONIC CRYSTAL CONFIGURATIONS

Lately, the research interests of EMC community have concentrated, apart from the aforementioned media layouts, on the potential applications of *periodic photonic crystals* [51–62]. Being able to interact with photons in a similar manner as the electrons in conventional crystals, these structures are typically composed of dielectric materials, although there are some rare implementations through metals. Depending on the inherent characteristics of the photonic crystal, it is possible to control the propagation of electromagnetic waves and create the so-called *photonic bandgaps (PBGs)*, namely, regions that permit the traveling of light inside the crystal at specific frequencies and directions of propagation. These features are mainly the type of periodicity (in one, two, or three dimensions), the relative dielectric permittivity of the crystal's medium, the differences between the properties of the media constructing the crystal, and the general geometry of the entire configuration. To realize the impact of these parameters on the function of a photonic crystal from an EMC point of view, the next paragraphs present some distinctive examples simulated by means of the FDTD method.

7.5.1   Propagation Through Dielectric-Rod Photonic Crystals

The first example investigates the flow of electromagnetic energy through a photonic crystal comprising a set of 18×9 infinitely long rods, as depicted in Figure 7.23a. Initially, all rods are meant to have a circular cross section with a radius of $r = 5$ mm and an intermediate distance among neighbors equal to $d = 13$ mm. For the purpose of the analysis, coefficient q is defined as the ratio of the areas occupied by the cross sections of the rods at a transverse cut of the crystal to the total area of the latter. Bearing in mind the periodicity of the structure, this ratio is, also, equal to $q = \pi(r/d)^2$, which in our case leads to $q = 0.4647$ (for $\varepsilon_r = 4$). The spatial increment of the FDTD lattice is $\Delta h = \Delta x = \Delta y = 0.5$ mm and $\Delta t = 0.9428$ ns, corresponding to the 80% of the maximum allowable time-step consistent with the Courant limit. Regarding the excitation of the system, the incident field is the modulated Gaussian pulse

$$E_{\text{inc}}|^n = 7e^{-16\left(\frac{n}{137}-1\right)^2} \cos\left[2\pi f_0(n - 137)\Delta t\right], \tag{7.37}$$

which carries the wide frequency content of 0–20 GHz. Finally, the domain's open ends are truncated by a six-cell PML absorber, appropriately tuned to attain high annihilation rates everywhere in the spectrum.

Despite the circular cross section of its rods, the photonic crystal can use other shapes, as well, such as rectangular or triangular ones. For the former, the side dimension is 9 mm ($q = 0.4793$), whereas for the latter the base is 12.5 mm and the height is 11.5 mm ($q = 0.4253$). Figure 7.23b illustrates the variation of the field spectrum for each of the cases. From the results, it can be deduced

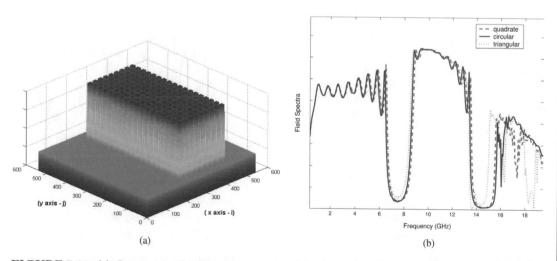

FIGURE 7.23: (a) Geometry of a photonic crystal comprising a set of 18×9 dielectric rods. (b) Field spectra of photonic crystals built by dielectric rods of different cross sections (circular, rectangular, and triangular).

that in the first PBG all geometries lead to similar patterns. However, in the second PBG, the triangular cross section appears to have a reduced (by almost 0.5 GHz) bandwidth and a defect mode (small peak) at 14 GHz. Actually, this particular geometry has the most complicated field spectrum, mainly attributed to the loss of axial symmetry in contrast to the rectangular and the cyclic ones. Moving to the influence of the rods' radius, Figure 7.24a shows the attenuation in the area of the PBG for several r values. As discerned, the increase of r decreases the attenuation of the crystal in the first PBG, whereas in the second one this situation appears to be smoother. It should be mentioned that the prior changes in r cause a respective ±0.5 GHz frequency shift of all PBGs.

Of equivalent importance is the choice of the effective dielectric constant ε_r, because it can allow for the fabrication of new EMC photonic devices, chiefly in the field of PCBs and radiating components. Thus, Figure 7.24b gives the variation of the field spectrum concerning three ε_r values, i.e., 2, 4, and 9. Evidently, the outcomes are very different, both in the number and the attenuation strength of the PBGs. In essence, the lower the selection of ε_r, the worse the attenuation performance of the crystal. Indeed, for $\varepsilon_r = 2$, only one and relatively weak PBG is acquired, rendering the layout inadequate for filtering implementations.

Another degree of freedom in the design of photonic crystals is the incorporation of deliberate defects, through the addition or removal of certain rod elements. This geometrical perturbation in the periodicity generates extra modes at certain frequencies in the PBG of the ideal crystal, which may be proven very instructive in the treatment of several electromagnetic problems. Herein, a number of 3, 9, and 18 rods have been omitted from the middle column of the crystal and the

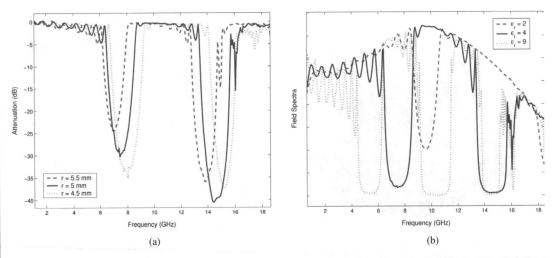

FIGURE 7.24: (a) Attenuation in the PBG as a function of frequency for diverse values of the rods' radius. (b) Variation of the field spectrum for various values of the rods' relative dielectric permittivity.

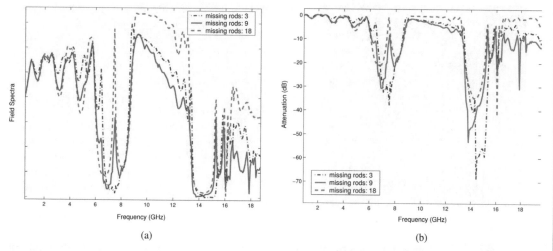

(a)   (b)

FIGURE 7.25: Generation of defect modes as a result of rod removal (periodicity disturbance) from the photonic crystal; 3, 9, and 18 rods have been, respectively, removed from the middle column of the structure. (b) The corresponding attenuation in the PBG due to the presence of defect modes.

simulation results for the field spectrum and the attenuation in the PBG are presented in Figure 7.25. A prompt conclusion reads that as the amount of the removed rods augments, two additional modes—at 7.5 and 15.3 GHz—inside the PBGs appear, which are amplified with the increase of the perturbation.

7.5.2 Waveguiding and Power Splitting via Photonic Crystals

Based on the notions and outcomes hitherto obtained, the insertion of defects is now applied in the design of more realistic EMC structures [52]. The first concept, depicted in Figure 7.26a, attempts to devise a "physical" waveguide by omitting a whole line of rods. Thus, the resulting element enables the propagation of specific waves through this path, because the rest of the space does not provide such a possibility. In this manner, waveguiding is conducted with the minimum possible losses at a large frequency range, even in the case of abrupt bends where optical fibers are proven insufficient. Figure 7.26b displays a snapshot of electric field intensity. Note that the incident pulse is practically reflected from the first column of the crystal except for the small opening of the removed line. Actually, the device operates as a waveguide between 10 and 11 GHz, a region located inside one of the crystal's PBGs.

The second application launches a more involved form of two different waveguide branches merged in one at a prefixed position in the photonic crystal (Figure 7.27a). Assuming that the layout retains its symmetry, the portions of power of the propagating waves at the two exits must be

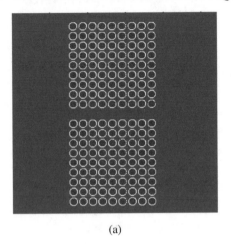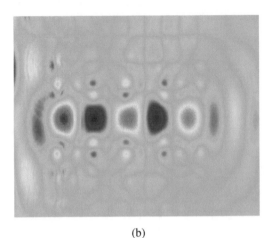

(a)   (b)

FIGURE 7.26: (a) A waveguiding structure formed on a 19 × 9-rod photonic crystal and (b) snapshot of its electric field intensity.

analogously divided, namely, the component operates as a power splitter. Figure 7.27b certifies these claims in terms of an indicative electric field snapshot that presents a satisfactory smoothness. Lastly, the power may be shared in a switching rationale among the two ports if the constitutive parameters of the media composing the crystal are locally modified.

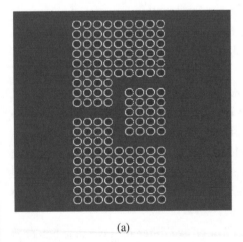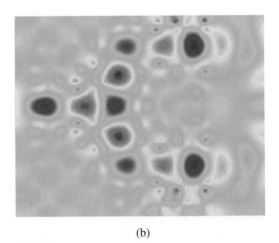

(a)   (b)

FIGURE 7.27: (a) A power splitter formed on a 19 × 9-rod photonic crystal and (b) snapshot of its electric field intensity.

REFERENCES

1. V. G. Veselago, "The electrodynamics of substances with simultaneously negative values of ε and μ," *Sov. Phys.-Usp.*, vol. 47, pp. 509–514, Jan.-Feb. 1968.

2. J. B. Pendry, A. J. Holden, D. J. Robbins, and W. J. Stewart, "Magnetism from conductors and enhanced nonlinear phenomena," *IEEE Trans. Microwave Theory. Tech.*, vol. 47, no. 11, pp. 2075–2084, Nov. 1999. doi:10.1109/22.798002

3. D. R. Smith, W. J. Padilla, D. C. Vier, S. C. Nemat-Nasser, and S. Schultz, "Composite medium with simultaneously negative permeability and permittivity," *Phys. Rev. Lett.*, vol. 84, pp. 4184–4189, May 2000. doi:10.1103/PhysRevLett.84.4184

4. C. Caloz and T. Itoh, *Electromagnetic Metamaterials: Transmission Line Theory and Microwave Applications. The Engineering Approach.* New York: John Wiley & Sons, 2006.

5. N. Engheta and R. W. Ziolkowski, *Electromagnetic Metamaterials: Physics and Engineering Explorations.* New York: John Wiley & Sons, 2006.

6. T. Weiland, R. Schuhmann, R. B. Greegor, C. G. Parazzoli, A. Vetter, D. R. Smith, D. C. Vier, and S. Schultz, "Ab initio numerical simulation of left-handed metamaterials: Comparison of calculations and experiments," *J. Appl. Phys.*, vol. 90, no. 10, pp. 5419–5424, Nov. 2001. doi:10.1063/1.1410881

7. P. Markós and C. M. Soukoulis, "Numerical studies of left-handed materials and arrays of split ring resonators," *Phys. Rev. E*, vol. 65, paper 036622, 2002. doi:10.1103/PhysRevE.65.036622

8. M. W. Feise, P. J. Bevelacqua, and J. B. Schneider, "Effects of surface waves on behavior of perfect lenses," *Phys. Rev. B, Condens. Matter*, vol. 66, paper 035113, Jul. 2002. doi:10.1103/PhysRevB.66.035113

9. G. V. Eleftheriades, O. Siddiqui, and A. K. Iyer, "Transmission line models for negative refractive index media and associated implementations without excess resonators," *IEEE Microwave Wireless Compon. Lett.*, vol. 13, no. 2, pp. 51–53, Feb. 2003.

10. S. A. Cummer, "Simulated causal subwavelength focusing by a negative refractive index slab," *Appl. Phys. Lett.*, vol. 82, no. 10, pp. 1503–1505, Mar. 2003. doi:10.1063/1.1554778

11. R. W. Ziolkowski, "Pulsed and CW Gaussian beam interactions with double negative metamaterial slabs," *Opt. Express*, vol. 11, pp. 662–673, Apr. 2003.

12. C. L. Holloway, E. F. Kuester, J. Baker-Jarvis, and P. Kabos, "A double negative (DNG) composite medium composed of magnetodielectric spherical particles embedded in a matrix," *IEEE Trans. Antennas Propagat.*, vol. 51, no. 10, pp. 2596–2603, Oct. 2003. doi:10.1109/TAP.2003.817563

13. R. Marqués, F. Mesa, J. Martel, and F. Medina, "Comparative analysis of edge- and broadside-coupled split ring resonators for metamaterial: Design, theory and experiments," *IEEE Trans. Antennas Propagat.*, vol. 51, no. 10, pp. 2572–2581, Oct. 2003. doi:10.1109/TAP.2003.817562

14. A. Alù and N. Engheta, "Pairing and epsilon-negative slab with a μ-negative slab: Resonance, tunneling and transparency," *IEEE Trans. Antennas Propagat.*, vol. 51, no. 10, pp. 2558–2571, Oct. 2003. doi:10.1109/TAP.2003.817553

15. M. K. Kärkkäinen, S. A. Tretyakov, S. I. Maslovski, and P. A. Belov, "A numerical study of evanescent fields in backward-wave slabs," *J. Phys. Condens. Matter*, vol. 20, paper 0302407, 2003.

16. L. Chen, S. He, and L. Shen, "Finite-size effects of a left-handed material slab on the image quality," *Phys. Rev. Lett.*, vol. 92, no. 10, paper 107404, Mar. 2004. doi:10.1103/PhysRevLett.92.107404

17. P. Baccarelli, P. Burghignoli, F. Frezza, A. Galli, P. Lampariello, G. Lovat, and S. Paulotto, "Effects of leaky-wave propagation in metamaterial grounded slabs excited by a dipole source," *IEEE Trans. Microwave Theory Tech.*, vol. 53, no. 1, pp. 32–44, Jan. 2005.

18. P. P. M. So, H. Du, and W. J. R. Hoefer, "Modeling of metamaterials with negative refractive index using 2D-shunt and 3D-SCN TLM networks," *IEEE Trans. Microwave Theory Tech.*, vol. 53, no. 4, pp. 1496–1505, Apr. 2005.

19. T. M. Grzegorczyk, C. D. Moss, J. Lu, X. Chen, J. Pacheco Jr., and J. A. Kong, "Properties of left-handed metamaterials: Transmission, backward phase, negative refraction, and focusing," *IEEE Trans. Microwave Theory Tech.*, vol. 53, no. 9, pp. 2956–2967, Sep. 2005.

20. C. L. Holloway, M. A. Mohamed, E. F. Kuester, and A. Dienstfrey, "Reflection and transmission properties of a metafilm: With an application to a controllable surface composed of resonant particles," *IEEE Trans. Electromagn. Compat.*, vol. 47, no. 4, pp. 853–864, Nov. 2005.

21. I. A. Eshrah, A. A. Kishk, A. B. Yakovlev, and A. W. Glisson, "Spectral analysis of left-handed rectangular waveguides with dielectric-filed corrugations," *IEEE Trans. Antennas Propagat.*, vol. 53, no. 11, pp. 3673–3683, Nov. 2005.

22. A. Grbic and G. V. Eleftheriades, "Negative refraction, growing evanescent waves, and sub-diffraction imaging in loaded transmission-line metamaterials," *IEEE Trans. Microwave Theory Tech.*, vol. 51, no. 12, pp. 2297–2305, Dec. 2005.

23. M. Gil, J. Bonache, I. Gil, J. García-García, and F. Martín, "On the transmission properties of left-handed micro-strip lines implemented by complementary split rings resonators," *Int. J. Numer. Model.*, vol. 19, no. 2, pp. 87–103, Mar.-Apr. 2006. doi:10.1002/jnm.601

24. N. V. Kantartzis and T. D. Tsiboukis, "Rigorous ADI-FDTD analysis of left-handed metamaterials in optimally-designed EMC applications," *COMPEL*, vol. 15, no. 3, pp. 677–690, 2006. doi:10.1108/03321640610666844

25. D. L. Sounas, N. V. Kantartzis, and T. D. Tsiboukis, "Focusing efficiency analysis and performance optimization of arbitrarily-sized DNG metamaterial slabs with loses," *IEEE Trans. Microwave Theory Tech.*, vol. 54, no. 12, p. 4111–4121, Dec. 2006.

26. H. Cory, Y. Lee, Y. Hao, and C. Parini, "Use of conjugate dielectric and metamaterial slabs as radomes," *IET Microwave Antennas Propagat.*, vol. 1, no. 1, pp. 137–143, 2007. doi:10.1049/iet-map:20050306

27. A. Sihvola, "Metamaterials in electromagnetics," *Metamaterials*, vol. 1, no. 1, pp. 2–11, 2007. doi:10.1016/j.metmat.2007.02.003

28. N. V. Kantartzis, D. L. Sounas, C. S. Antonopoulos, and T. D. Tsiboukis, "A wideband ADI-FDTD algorithm for the design of double negative metamaterial-based waveguides and antenna structures," *IEEE Trans. Magn.*, vol. 43, no. 4, pp. 1329–1332, Apr. 2007.

29. H. Mosallaei, "FDTD-PLRC technique for modeling of anisotropic-dispersive media and metamaterial devices," *IEEE Trans. Electromagn. Compat.*, vol. 49, no. 3, pp. 649–660, Aug. 2007.

30. D. L. Sounas, N. V. Kantartzis, and T. D. Tsiboukis, "Temporal characteristics of resonant surface polaritons in superlensing planar double-negative slabs: Development of analytical schemes and numerical models," *Phys. Rev. E*, vol. 75, no. 4, paper 046606, 2007. doi:10.1103/PhysRevE.76.046606

31. M. Y. Koledintseva, J. L. Drewniak, D. J. Pommerenke, G. Antonini, A. Orlandi, and K. N. Rozanov, "Wide-band Lorentzian media in the FDTD algorithm," *IEEE Trans. Electromagn. Compat.*, vol. 47, no. 2, pp. 392–398, May 2005. doi:10.1109/TEMC.2005.847406

32. I. Lindell, A. Sihvola, S. Tretyakov, A. Viitanen, *Electromagnetic Waves in Chiral and Bi-Isotropic Media*. Boston, MA: Artech House, 1994.

33. S. González García, I. Villó Pérez, R. Gómez Martín, S. García Olmedo, "Extension of Berenger's PML for bi-isotropic media," *IEEE Microwave Guided Wave Lett.*, vol. 8, no. 9, pp. 297–299, Sept. 1998. doi:10.1109/75.720460

34. A. Serdyukov, I. Semchenko, S. Tretyakov, and A. Sihvola, *Electromagnetics of Bi-Anisotropic Materials: Theory and Applications*. Amsterdam, The Netherlands: Gordon and Breach Science Publishers, 2001.

35. A. Akyurtlu and D. Werner, "Bi-FDTD: A novel finite-difference time-domain formulation for modeling wave propagation in bi-isotropic media," *IEEE Trans Antennas Propagat.*, vol. 52, no. 2, pp. 416–425, Feb. 2004. doi:10.1109/TAP.2004.823956

36. C. P. Neo and V. K. Varadan, "Optimization of carbon fiber composite for microwave absorber," *IEEE Trans. Electromagn. Compat.*, vol. 46, no. 1, pp. 102–106, Feb. 2004.

37. A. Grande, I. Barba, A. Cabeceira, J. Represa, P. So, W. Hoefer, "FDTD modeling of transient microwave signals in dispersive and lossy bi-isotropic media," *IEEE Trans Microwave Theory Tech.*, vol. 52, no. 3, pp. 773–784, Mar. 2004. doi:10.1109/TMTT.2004.823537

38. C.-N. Chiu and I.-T. Chiang, "Transient reflection properties of a dispersive lossy bi-isotropic slab with an anisotropic laminated composite backing," *IEEE Trans. Electromagn. Compat.*, vol. 47, no. 4, pp. 845–852, Nov. 2005.

39. A. Semichaevsky, A. Akyurtlu, D. Werner, and M. Bray, "Novel BI-FDTD approach for the analysis of chiral cylindrical spheres," *IEEE Trans Antennas Propagat.*, vol. 54, no. 3, pp. 925–932, Mar. 2006. doi:10.1109/TAP.2006.869898

40. D. X. Wang, E. K. N. Yung, R. S. Chen, and P. Y. Lau, "An efficient volume integral equation solution to EM scattering by complex bodies with inhomogeneous bi-isotropy," *IEEE Trans. Antennas Propagat.*, vol. 55, no. 7, pp. 1970–1980, Jul. 2007.

41. N. Wongkasem, A. Akyurtlu, K. A. Marx, Q. Dong, J. Li, and D. W. Goodhue, "Development of chiral negative refractive index metamaterials for the terahertz frequency regime," *IEEE Trans. Antennas Propagat.*, vol. 55, no. 11, pp. 3052–3062, Nov. 2007.

42. C.-W. Qiu, S. Zouhdi, and A. Razek, "Modified spherical wave functions with anisotropy ratio: Application to the analysis of scattering by multilayered anisotropic shells," *IEEE Trans. Antennas Propagat.*, vol. 55, no. 12, pp. 3515–3523, Dec. 2007.

43. G. Bouzianas, N. V. Kantartzis, and T. D. Tsiboukis, "Development of accuracy-enhanced time-domain schemes for bi-isotropic media and chiral metamaterials," *COMPEL*, vol. 4, 2008.

44. M. Sarto, F. Sarto, M. C. Larciprete, M. Scalora, M. D'Amore, C. Sibilia, and M. Bertolotti, "Nanotechnology of transparent metals for radiofrequency electromagnetic shielding," *IEEE Trans. Electromagn. Compat.*, vol. 45, no. 4, pp. 586–594, Nov. 2003.

45. D. P. Fromm, A. Sundaramurthy, P. J. Schuck, G. S. Kino, and W. E. Moerner, "Gap-dependent optical coupling of single 'bowtie' nanoantennas resonant in the visible spectrum," *Nano Lett.*, vol. 4, pp. 957–961, 2004.

46. C. Oubre and P. Nordlander, "Optical properties of metallodielectric nanostructures calculated using the finite difference time domain method," *J. Phys. Chem. B.*, vol. 108, pp. 17740–17747, 2004.

47. M. S. Sarto, R. L. Voti, F. Sarto, and M. C. Larciprete, "Nanolayered lightweight flexible shields with multidirectional optical transparency," *IEEE Trans. Electromagn. Compat.*, vol. 47, no. 3, pp. 602–611, Aug. 2005.

48. Y. Zhao and Y. Hao, "Finite-difference time-domain study of guided modes in nano-plasmonic waveguides," *IEEE Trans. Antennas Propagat.*, vol. 55, no. 11, pp. 3070–3077, Nov. 2007.

49. K. L. Wong, *Compact and Broadband Microstrip Antennas*. Piscataway, NJ: Wiley Interscience, 2002. doi:10.1002/0471221112.ch3

50. K. P. Prokopidis and T. D. Tsiboukis, "The effect of substrate dispersion on the operation of square microstrip antennas," *IEEE Trans. Magn.*, vol. 42, no. 4, pp. 603–606, Apr. 2006.

51. E. Yablonovitch, "Inhibited spontaneous emission in solid-state physics and electronics," *Phys. Rev. Lett.*, vol. 58, pp. 2059–2062, 1987. doi:10.1103/PhysRevLett.58.2059

52. J. D. Joannopoulos, R. D. Meade, and J. N. Winn, *Photonic Crystals: Molding the Flow of Light*. New York, NY: Princeton University Press, 1995.

53. D. Sievenpiper, *High-impedance electromagnetic surfaces*, Ph.D. thesis, UCLA, 1999.

54. M. Bayindir, B. Temelkuran, and E. Ozbay, "Propagation of photons by hopping: a waveguiding mechanism through localised coupled-cavities in three-dimensional photonic crystals," *Phys. Rev. B*, vol. 61, R11855, 2000. doi:10.1103/PhysRevB.61.R11855

55. S. Mingaleev and Y. Kivshar, "Non-linear photonic crystals: towards all optical technologies," *Opt Photonics News*, pp. 48–62, Jul. 2002.

56. H. Takeda and K. Yoshino, "Tunable light propagation in Y-shaped waveguides in two-dimensional photonic crystals utilizing liquid crystals as linear defects," *Phys. Rev. B*, vol. 61, paper 073106, 2003. doi:10.1103/PhysRevB.67.073106

57. H. Kubota, S. Kawanishi, S. Koyanagi, M. Tanaka, and S. Yamaguchi, "Absolutely single polarization photonic crystal fiber," *IEEE Photon. Technol. Lett.*, vol. 16, no. 1, pp. 182–184, Jan. 2004. doi:10.1109/LPT.2003.819415

58. M. W. Haakestad, T. T. Alkeskjold, M. Nielsen, L. Scolari, J. Riishede, H. E. Engan, and A. Bjarklev, "Electrically tunable photonic bandgap guidance in a liquid-crystal-filled photonic crystal fiber," *IEEE Photon. Technol. Lett.*, vol. 17, no. 4, pp. 819–821, Apr. 2005.

59. E. P. Kosmidou, Em. E. Kriezis, and T. D. Tsiboukis, "Analysis of tunable photonic crystal devices comprising liquid crystal materials as defects," *IEEE J. Quant. Electron.*, vol. 41, no. 5, pp. 657–665, 2005.

60. D. C. Zografopoulos, Em. E. Kriezis, and T. D. Tsiboukis, "Tunable highly birefringent bandgap-guiding liquid-crystal microstructured fibers," *IEEE J. Lightwave Technol.*, vol. 24, no. 9, pp. 3427–3432, Sept. 2006,

61. E. P. Kosmidou, Em. E. Kriezis, and T. D. Tsiboukis, "Analysis of tunable photonic crystal directional couplers," *J. Appl. Phys.*, vol. 100, pp. 1–9, 2006.

62. D. C. Zografopoulos, Em. E. Kriezis, and T. D. Tsiboukis, "Photonic crystal-liquid crystal fibers for single-polarization or high-birefringence guidance," *Opt. Express*, vol. 13, no. 2, pp. 914–925, 2006.

·   ·   ·   ·

Authors Biographies

Nikolaos V. Kantartzis received the Diploma degree and Ph.D. degree in electrical and computer engineering from the Aristotle University of Thessaloniki (AUTH), Thessaloniki, Greece, in 1994 and 1999, respectively.

In 1999, he joined the Applied and Computational Electromagnetic Laboratory, Department of Electrical and Computer Engineering, AUTH, as a Postdoctoral Research Fellow. He coauthored *Higher-Order FDTD Schemes for Waveguide and Antenna Structures* (Morgan & Claypool Publishers, 2006). He has authored or coauthored several refereed journal papers in the area of electromagnetic compatibility (EMC), computational electromagnetics and especially higher order finite-difference time-domain methods, perfectly matched layers (PML), and vector finite elements. His main research interests include EMC modeling, time- and frequency-domain algorithms, double-negative (DNG) metamaterials, waveguides, and antenna structures.

Theodoros D. Tsiboukis received the Diploma degree in electrical and mechanical engineering from the National Technical University of Athens, Athens, Greece, in 1971, and the Ph.D. degree from the Aristotle University of Thessaloniki (AUTH), Thessaloniki, Greece, in 1981.

From 1981 to 1982, he was with the Electrical Engineering Department, University of Southampton, Southampton, U.K., as a Senior Research Fellow. Since 1982, he has been with the Department of Electrical and Computer Engineering (DECE), AUTH, where he is currently a Professor. He has served in numerous administrative positions including Director of the Division of Telecommunications, DECE (1993–1997) and Chairman, DECE (1997–2001). He is also the Head of the Advanced and Computational Electromagnetics Laboratory, DECE. He has authored or coauthored eight books and textbooks including *Higher-Order FDTD Schemes for Waveguide and Antenna Structures* (Morgan & Claypool Publishers, 2006). He has authored or coauthored over 135 refereed journal papers and more than 100 international conference papers. He was the Guest Editor of a special issue of the *International Journal of Theoretical Electrotechnics* (1996). His main research interests include electromagnetic-field analysis by energy methods, computational electromagnetics (finite-element method (FEM), boundary-element method (BEM), vector finite elements, method of moments (MoM), finite-difference time-domain (FDTD) method, alternating-direction implicit

(ADI)-FDTD method, integral equations, and absorbing boundary conditions), metamaterials, photonic crystals, and EMC problems.

Prof. Tsiboukis is a member of various societies, associations, chambers, and institutions. He was the chairman of the local organizing committee of the 8th International Symposium on Theoretical Electrical Engineering (1995). He has been the recipient of several awards and distinctions.

Printed in the United States
by Baker & Taylor Publisher Services